ISO 14001
Implementation
Manual

ISO 14001 Implementation Manual

Gayle Woodside

Patrick Aurrichio

Jeanne Yturri

McGraw-Hill

New York   San Francisco   Washington, D.C.   Auckland   Bogotá
Caracas   Lisbon   London   Madrid   Mexico City   Milan
Montreal   New Delhi   San Juan   Singapore
Sydney   Tokyo   Toronto

Library of Congress Cataloging-in-Publication Data

Woodside, Gayle.
    ISO 14001 implementation manual / Gayle Woodside, Patrick
Aurrichio, Jeanne Yturri.
       p.   cm.
    Includes index.
    ISBN 0-07-071852-0
    1. ISO 14000 Series Standards.   2. Production management—
Environmental aspects.   I. Aurrichio, Patrick.   II. Yturri,
Jeanne.   III. Title.
TS155.7.W66   1998
658.4'08—dc21   98-11899
                                            CIP

McGraw-Hill

*A Division of The **McGraw·Hill** Companies*

8 9 0   IBT/IBT   0 1 0 9 8 7

ISBN 0-07-071852-0

*The sponsoring editor for this book was Robert Esposito, the editing supervisor
was Curt Berkowitz, and the production supervisor was Pamela Pelton. It was
set in Palatino by Dina John of McGraw-Hill's Professional Book Group com-
position unit.*

Printed and bound by R. R. Donnelley & Sons Company.

This book is printed on recycled, acid-free paper containing a
minimum of 50% recycled de-inked fiber.

Contents

Part 2. Planning

Part 3. Implementation and Operation

Foreword

It is quite evident, only one year after its inauguration, that the ISO 14000 series of environmental management standards will have a monumental impact on worldwide efforts and strategies to protect the environment. A long list of governmental agencies, standards bodies, and industry associations have been busy mobilizing resources to acquire knowledge and engage expertise for building their national infrastructures to implement the new standards—especially the ISO 14001 Environmental Management System. Already, at the end of 1997, some 2500 facilities have been registered as conforming to the requirements of ISO 14001. (An additional 1200 facilities have been registered as conforming to the European Union's Eco-Management and Audit Scheme.)

Why all this interest and excitement about an environmental management standard that does not set performance levels, does not prescribe technology solutions, and makes no attempt to define how clean is clean or what is environmentally acceptable? Mostly, I would speculate, because it lays out a practical approach for addressing environmental concerns and because it answers the needs of sincere individuals—in government, industry, and commerce—to come to grips with their environmental obligations. It allows them to initiate appropriate and acceptable action that can lead to tangible results without compromising the goals and progress of their enterprises or their nations. It lets them implement a management framework for achieving continual improvement that is in every respect very flexible and adaptable to a very wide spectrum of situations, economics, organizations, and national circumstnaces. From that perspective, ISO 14001 may be the first tool that has been put at our disposal to move them closer to the desired, but often elusive, path of sustainable development. More specifically, it achieves many desirable ends, some of which are noted below:

- *Builds Capacity.* Environmental management systems place the emphasis on capacity building, awareness raising, operational integration, and behavior modification rather than simply focusing on legal requirements. The fact is that, in most organizations today, only a very small minority of employees are involved with the environmental concerns or legal requirements and obligations of their enterprise. Having been left out, the greater portion of the workforce is mostly oblivious to the environmental dimension of their work. By integrating this larger segment of the employee population into the organization's strategies—as they address operations programs, regulatory requirements, control techniques, remediation activities, and prevention discipline—environmental management systems are able to engage larger portions of available human energy, knowledge, and creativity to effect improvements in operational efficiency, regulatory compliance, and protection of the environment. A similar approach has already been proven for quality management where the advent of ISO 9000 in 1987 induced revolutionary changes and results in quality management. There is every reason to believe that comparable structural change in the environmental sphere will also bring notable improvements in the coming years for environmental protection and performance.

- *Increases Industrial Efficiencies.* Environmental management systems primarily address the inefficiency of industrial operations that are related to the use of materials, substances, energy, water, and land. Environmental protection is not only about protecting human and ecological health—important as these are—but also about competitiveness, profitability, and sustainability, which are greatly improved by minimizing waste and the misuse of resources. Here too, through environmental management systems, industry can instill employee attitudes to avoid such waste and misuse as part of an ethic that promotes both respect for the natural world and for the welfare of the enterprise.

- *Promotes Trade.* Just like the ISO 9000 Quality Management Standards, the ISO 14000 standards are becoming a condition of international trade. International enterprises want to demonstrate their interest in participating in world trade as responsible, aware, and considerate players willing to adopt and adhere to world standards. International trade relies on confidence, trust, and shared views. Most enterprises are increasingly dependent on foreign trade for their growth and prosperity. Adopting and implementing the ISO 14000 standards will send a strong signal to the world community that an entity wants to participate in the global economy by following global norms.

■ *Attracts Foreign Capital.* Many organizations are eager to attract foreign capital and investment. To a significant extent, this is dependent on how foreign investors and entrepreneurs gauge the risk of making such investments. Environmental risks can be, and in many cases have been, deterrents to investors. Enterprises are increasingly aware of such obstacles and want to demonstrate their resolve to minimize such risks through the implementation of environmental management systems.

■ *Achieves Global Recognition.* Industry is also aware that global recognition for environmentally responsible behavior cannot always be achieved by following national regulations alone. There is now a growing global expectation to demonstrate commitment to structural change that promotes continual improvements that ultimately result in the least environmental impacts. Gaining recognition as a company that performs to those levels of responsible behavior can best be achieved through ISO 14001 registration. For this reason, implementation of ISO 14001 can be particularly important for enterprises that are seeking to integrate into the global economy.

■ *Improves Compliance Assurance.* Organizations that implement ISO 14001 environmental management systems acquire the discipline and confidence to continually improve their environmental performance, increase the reliability and consistency of their compliance systems, and diminish the potential for disastrous occurrences. Such organizations can be counted on to a greater degree to reach compliance with national regulations and maintain it with greater consistency.

■ *Enhances Environmental Security and Global Stabilization.* ISO 14001 implementation has been recognized by various national defense establishments as an important element in promoting "environmental security." Environmental management systems build the infrastructures that lead to environmental security for individual countries and for the world community. There is no better way to channel assistance to developing countries than to help build their infrastructures for sustained growth and stability. Environmental management systems can be the pivotal component in efforts to achieve environmental security and stability in such countries.

- *May Impact Climate Change Emissions.* ISO 14001 can provide the foundation for industry's efforts to reduce emissions of gases that contribute to climate change. Environmental management is fundamentally a discipline that builds institutional awareness and promotes management decisions that reduce environmental impacts. Given the high visibility and global focus on climate change, all organizations that implement environmental management systems will find it imperative to address greenhouse gases within their systems. ISO 14001 has the potential to become a major contributor in countries' strategies to satisfy their international commitments on this issue.

This book comes at a very opportune time in the life of ISO 14001. While many volumes have been published over the last two years that adequately describe ISO 14001, this one alone helps us to implement that standard. It is precisely what is needed at this juncture. For while many large and sophisticated organizations have led the way onto the implementation trail, the vast majority of smaller establishments are just now pondering their course of action and can use all the assistance they can get. The authors make excellent use of their own efforts in implementing ISO 14001, and there is no better teacher than actual "hands on" experience. While ISO 14001 is a relatively easy read, the interpretation and approach to a number of its requirements are not always as straightforward as they appear at first. This volume will serve as a handy and inexpensive reference for small- and medium-sized enterprises that wish to avoid the expense of external consultants. And this, as it turns out, is exactly what the drafters of ISO 14001 intended—to give organizations a system that they could implement themselves with only the assistance and participation of their employees. The authors, to our relief and gratitude, have now made good on that promise.

Joe Cascio, Chairman
U.S. Technical Advisory Group
to ISO TC 207

Preface

When ISO 14001 was being drafted in the early and mid-1990's, some experts raised doubts concerning the standard's potential usefulness in the workplace. Now, organizations of all sizes and complexity that have implemented the standard have found that it provides an effective, comprehensive framework for environmental management. The environmental management system laid out in the standard provides a holistic approach that weaves the responsibility for environmental care into the fabric of the organization at all levels. Although the focus of the standard is the environmental management system and not environmental performance, once implemented the management system provides the structure necessary to achieve continual improvement in all areas of environmental management, including environmental performance.

Implementing ISO 14001 is a demanding task. It requires top management commitment; an understanding of environmental aspects of the organization's activities, products, and services; the setting and tracking of objectives and targets; employee training and awareness; and much more. Our goal in writing *ISO 14001 Implementation Manual* is to provide complete and detailed implementation guidance for each element of the standard. Using a hypothetical company that is implementing the standard, we have provided the reader with what we hope is a practical method for meeting ISO 14001 requirements. The book contains examples of a gap analysis, procedures required by the standard, an environmental management system audit methodology, and training overheads. In addition, we discuss the intent of each element of the standard and provide numerous checklists, figures, and tables to add focus to key points.

Implementing ISO 14001 is a journey, and putting all required elements of the standard in place is only the beginning. Commitments to pollution prevention, continual improvement, and compliance with legal requirements are ongoing. As environmental objectives and targets are achieved, new ones are set. As employees become better aware of the consequences of their actions upon the environment, accidents decrease and performance increases. We feel from our personal experiences with ISO 14001 that once embarked upon, the journey which uses the standard as its navigational compass will ultimately benefit the organization, earth, and humanity.

Gayle Woodside

Patrick Aurrichio

Jeanne Yturri

ISO 14001
Implementation
Manual

PART 1

Background, General Requirements, and Environmental Policy

1
ISO 14001: An Introduction

General Overview

The International Organization for Standardization (ISO) has been a leader in the international technical arena for a half decade. "ISO" is commonly used when referring to the organization and its standards, but the term "iso" is actually a Greek word meaning "equal." Technically speaking, ISO is primarily a body for developing products and safety standards; however, in the late 1980s, it journeyed off its traditional path by developing a series of quality management standards—the ISO 9000 series. These quality management standards are process standards (that is, they specify a process and not an end goal), the first of their kind for ISO. The ISO 9000 series brought world recognition to the organization as a leader in developing these types of standards. Further, although all ISO standards are termed "voluntary," in some countries registration to the appropriate ISO 9000 standard(s) has become a requirement for trade.

With the resounding success of the ISO 9000 series, in mid-1991 ISO once again embarked on a new journey—to environmental management standards. Like the ISO 9000 series, these standards are process standards. They are not intended to set environmental performance goals, but rather to specify the elements of a management system that provides a framework for an organization to develop and maintain a reliable process that consistently meets environmental obligations and commitments.

Although the focus of this book is the ISO 14001 standard—termed "Environmental Management Systems: Specification with Guidance for Use"—it is appropriate to discuss briefly the other standards that have been issued or are being developed under the ISO 14000 umbrella.

Simply put, there are five types of standards, a guide, and a set of terms and definitions which tie the verbiage of the standards together.[1] Figure 1-1 illustrates the ISO family of standards and the interconnection between them. As shown in the figure, there are two discrete types of standards—organization evaluation and product evaluation. There are three organization-evaluation standards—environmental management systems (EMS) specification (ISO 14001), environmental auditing (EA), and environmental performance evaluation (EPE). The latter two standards are actually interrelated with ISO 14001 in that they provide guidance (but not requirements) for implementation of an EMS audit and for monitoring and measurement of key characteristics—two important elements of ISO 14001. There are two product-evaluation standards and a guide, the standards being product labeling and life cycle assessment (LCA). Environmental aspects in product standards (EAPS) was formerly a standard but, once drafted, was redefined as a guide.

Table 1-1 details the names and purposes of the key documents in the ISO 14000 family. These standards were crafted (and in some cases are still being crafted) by experts around the world. In addition, ISO is always considering the development of new environmental management standards that support the existing standards or that define new areas.

[1]Additional information on the background of the ISO 14000 standards can be found in *ISO 14000 Guide*, McGraw-Hill, 1996.

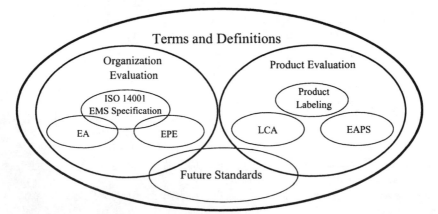

Figure 1-1. ISO 14000 family of standards.

Table 1-1. ISO 14000 Family of Standards

Environmental Management Systems (EMS) Standards

ISO 14001—Environmental Management Systems Specification This standard lays out elements of an EMS; it applies to organizations of all sizes, and those who seek registration to the standard must conform to all requirements delineated within.

ISO 14004—Guidance Document for EMS This document provides guidance on setting up an EMS; it is most useful for those organizations with immature systems.

Environmental Auditing (EA) Standards

ISO 14010—General Principles of Environmental Auditing This document provides general principles of environmental auditing; it is meant to apply to all types of environmental auditing and covers topics such as objectivity, independence, competence, due professional care, and audit criteria.

ISO 14011—Auditing Procedures—Auditing of Environmental Management Systems This document provides guidance for the EMS audit; although its use is not required, it provides pertinent information on how to develop an audit plan and conduct of an EMS audit.

ISO 14012—Auditor Qualification Criteria This document sets forth guidance for qualification criteria for internal and external environmental auditors who perform EMS audits.

Environmental Labels and Declarations Standards

ISO 14020—Principles of All Environmental Labeling This document provides guidance on the goals and principles that should be consistently incorporated into all types of environmental labeling programs.

ISO 14021—Self-Declaration of Environmental Claims—Terms and Definitions This document is meant to apply to manufacturers who are declaring that their product has an environmental attribute (i.e., it is recyclable, energy-efficient, etc.).

ISO 14022—Symbols This document addresses environmental labeling symbols.

ISO 14023—Testing and Verification This document addresses testing and verification of environmental labeling claims.

ISO 14024—Practitioner Programs: Guiding Principles and Procedures for Multiple Criteria (Type 1) This document lays out the principles and protocols for environmental labeling programs.

Environmental Performance Evaluation (EPE)

ISO 14031—Environmental Performance Evaluation This document addresses performance indicators for operations in terms of emissions and efficiencies of processes; it also addresses performance indicators for the environment itself.

Table 1-1. ISO 14000 Family of Standards (*Continued*)

Life-Cycle Assessment (LCA) Standards

ISO 14040—Principles and Framework   This document provides a clear overview of the practice, applications, and limitations of LCA to a broad range of potential LCA users.

ISO 14041—Goal/Scope Definitions and Inventory Analysis   This document describes special requirements and guidelines for the preparation, conduct, and critical review of the life-cycle inventory analysis.

ISO 14042—Impact Assessment   This document provides guidance on the impact phase of the life-cycle assessment.

ISO 14043—Interpretations   This document provides guidance on how to interpret the life-cycle assessment of impacts.

Terms and Definitions

ISO 14050—Terms and Definitions   This document harmonizes the terms and definitions used in the various ISO 14000 documents.

Environmental Aspects in Product Standards (EAPS)

Guide 64 (Formerly ISO 14060)—Guide for Environmental Aspects in Product Standards
This guide helps standards writers avoid inserting specifications in product standards that could turn out to be environmentally detrimental.

ISO 14001: Environmental Management Systems Specification

ISO 14001, which was finalized and issued as a first edition Sept. 1, 1996, is the most widely recognized environmental management standard. It is a specification standard, which means that organizations which conform to its requirements can become registered to the standard.[2] ISO 14001 was written as a consensus standard with nearly 50 countries participating. The standard is applicable to all types and sizes of organizations, and it accommodates diverse geographical, cultural, and social conditions. It can be applied to all parts or any single part of an organization and/or its activities, products, and services. A basic model of the standard is depicted in Fig. 1-2.

---

[2] Outside the United States, the term "certified" is used instead of "registered"; although the two terms are synonymous, the authors will use the term "registered" throughout the book for consistency.

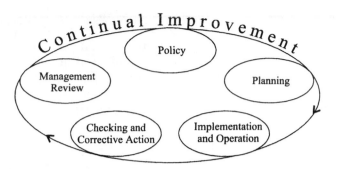

Figure 1-2. Environmental management system model.

The ISO 14001 standard is organized as follows:

- *Introduction.* This nonmandatory section sets the tone and context for implementation of the standard. In addition to background information about applicability and use of the standard, it provides a very basic model of the framework that the standard sets, which includes the five major elements and the concept of continual improvement. This model is presented in Fig. 1-2.

- *Scope (Section 1).* This section specifies applicability and introduces Annex A as nonmandatory guidance.

- *Normative references (Section 2).* None are listed at present.

- *Definitions (Section 3).* Thirteen definitions that apply to the standard are presented. Examples include definitions of environmental aspect, continual improvement, environmental performance, and interested party.

- *Environmental management system (EMS) requirements (Section 4).* This section details the requirements of the EMS to which an organization must conform if it wants to become registered to the standard.

- *Annex A.* This nonmandatory (informative) annex provides guidance on the EMS requirements section so that misinterpretation of the standard can be avoided.

- *Annex B.* This second nonmandatory (informative) annex provides two tables that identify the links and broad technical correspondences between ISO 14001 and ISO 9001.

- *Annex C.* This final nonmandatory (informative) annex provides a bibliography of ISO 9000 standards, environmental audit guidelines, and an EMS guidance document.

There are five major elements of the standard, as shown in Fig. 1-2—policy, planning, implementation and operation, checking and corrective action, and management review. These elements interact with each other to form the framework of an integrated, systematic approach to environmental management, with the ultimate result being continual improvement of the overall system. A more detailed illustration of how the key elements and their subelements interact is shown in Fig. 1-3.

As the figure shows, the five elements actually build upon each other, with environmental policy being the first tier in the structure—the base

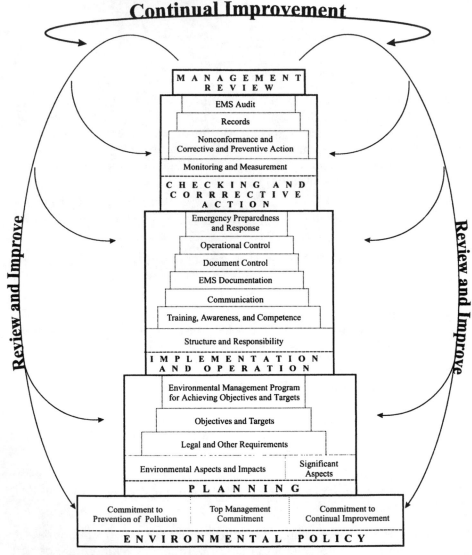

Figure 1-3. Key elements and subelements of ISO 14001.

Benefits of Implementing ISO 14001

✓ Positions the organization in the marketplace as an environmental leader

✓ Provides a framework for establishing an integrated approach to environmental management that is system-dependent and not person-dependent

✓ Promotes a positive image in the community and with governmental agencies

✓ Provides a systematic, repeatable process to achieve continual improvement of the organization's environmental management system, thereby promoting improvement of environmental performance

✓ Supports integration of relevant levels and functions into the environmental management system

✓ Provides an internationally recognized standard for the organization to use to demonstrate to employees and interested parties its commitment to sound environmental management

---

Figure 1-4. The benefits of implementing ISO 14001.

which supports the entire framework of the EMS. The subelements within each of the key elements also build upon each other. Viewed as a structure, it is evident that all key elements and subelements are needed for a strong EMS. Any missing or weak part could effectively "topple" the system. Finally, the figure illustrates the concept of continual improvement as being improvement of any element of the EMS. With continual improvement of the EMS, the organization can expect to achieve, as a secondary benefit, continual improvement of environmental performance. Another outcome of implementing the requirements of ISO 14001 is the development of an EMS that is system-dependent and not person-dependent.

Although ISO 14001 is not meant to replace legal/regulatory schemes, it can serve to augment existing regulatory programs worldwide. Benefits of implementing this international standard are presented in Fig. 1-4.

The next chapters of the book discuss the sections of the ISO 14001 standard in sequential order. In addition to describing the first element of the standard, Chap. 2 provides a description of a hypothetical company—Quality Seat Belts, Inc. This hypothetical company will be used to provide examples of ways that the elements of ISO 14001 can be implemented at an international company.

References

Cascio, Joseph, Gayle Woodside, and Philip Mitchell, *ISO 14000 Guide: The New International Environmental Management Standards,* McGraw-Hill, New York, 1996.

2
Section 4.1: General Requirements

Introduction

The Intent of Section 4.1—General Requirements

Section 4.1 of ISO 14001 is brief and to the point. Essentially, the section requires an organization to establish and maintain an environmental management system (EMS) that conforms to the requirements detailed throughout the rest of Section 4—the mandatory sections of the standard. To understand this requirement, it is appropriate to review several definitions that pertain to the standard and the EMS.

First, what is an "organization"? The standard defines it as a "company, corporation, firm, authority or institution, or part or combination thereof, whether incorporated or not, public or private, that has its own function and administration" (ISO 14001, Section 3.12). This very broad and flexible definition essentially allows many organizational entities to align their EMS's to the requirements of ISO 14001. Examples of what could be considered "organizations" are presented in Fig. 2-1.

Another definition that is integral to the standard is "environmental management system" (ISO 14001, Section 3.5). The ISO 14001 standard defines this term as "the part of the overall management system that includes orga-

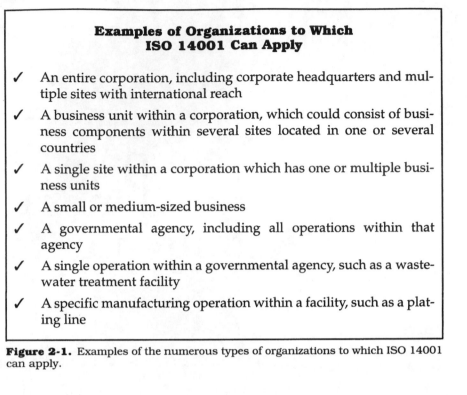

Figure 2-1. Examples of the numerous types of organizations to which ISO 14001 can apply.

nizational structure, planning activities, responsibilities, practices, procedures, processes, and resources for developing, implementing, achieving, reviewing, and maintaining the environmental policy." In short, it defines the system that is depicted in Fig. 1-3 in Chap. 1.

Further, the term "continual improvement" (ISO 14001, Section 3.1) warrants some discussion at this point. The writers of the standard were careful in crafting the verbiage of this definition, which reads "process of enhancing the environmental management system to achieve improvements in overall environmental performance in line with the organization's environmental policy." Thus "continual improvement" is not improvement of environmental performance per se, but rather improvement of the EMS, which then leads to improvement in environmental performance.

Other important definitions include "environmental aspect," "environmental impact," and "environmental management system audit." These terms will be discussed in the appropriate chapters encompassing these elements of the standard—Chap. 3 for "environmental aspects" and "environmental impacts," and Chap. 18 for "environmental management system audit."

Structuring an Environmental Management System

Although the improvement of environmental performance is not a prerequisite to obtaining registration to ISO 14001, implementation of the elements of the standard, as mentioned above, should likely result in such improvements. The requirements follow the "plan, do, check, and act" model of the ISO 9000 standard, which forms the basis of continual improvement of the system. The implementation of an EMS which conforms to ISO 14001 should allow the organization to accomplish the following management system objectives:

- Establish an environmental policy appropriate to the organization's activities, products, and services.
- Identify significant environmental aspects and impacts.
- Identify legal requirements.
- Set environmental objectives and targets and the program to accomplish them.
- Define structure and responsibility for elements of the EMS.
- Involve employees at relevant levels in the EMS process.
- Enhance communication with employees and external interested parties.
- Schedule audits of the EMS.
- Establish schedules for management review of the EMS for suitability, adequacy and effectiveness, and continual improvement of the EMS itself.

Quality Seat Belts, Inc.—A Hypothetical International Company That Is Implementing ISO 14001

In order to more fully understand how a company can conform to the requirements of ISO 14001, the authors have developed a hypothetical company called Quality Seat Belts, Inc. (QSBI). This fictitious company, which is described below, will be used to illustrate what is required within the various sections of the ISO 14001 standard.

Description of Products Manufactured by Quality Seat Belts, Inc.

Quality Seat Belts, Inc. (QSBI), established in 1967, is a fictitious company serving as a model for implementing an EMS that conforms to the requirements of ISO 14001. The mission of QSBI is to manufacture seat belts in a high-quality, cost-effective, environmentally conscious manner. QSBI manufactures seat belts for a variety of applications and customers including cars, trucks, vans, school busses, and airplanes. A secondary product made by QSBI, which was introduced in 1984, is fasteners. Examples of these products include fasteners used on car seats for children, fasteners for baby highchairs, and fasteners for luggage.

To manufacture the seat belt and fastener products, several types of metal and plastic components are produced. Textile materials used for the belts are purchased, although the dimensioning (cutting) and stitching is performed at QSBI. In addition, connection components of the belt (i.e., connecting brackets) are manufactured at QSBI.

Demographics of Quality Seat Belts, Inc.

One goal of QSBI is to provide jobs in markets where QSBI sells its products. Thus, with growth over several decades, QSBI has expanded its operations significantly and has worldwide operations, as shown in Fig. 2-2.

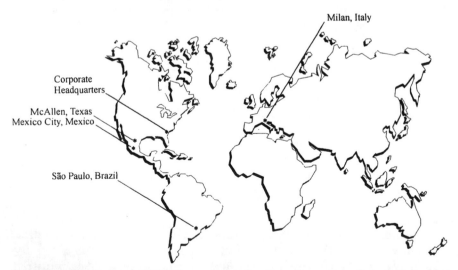

Figure 2-2. Locations of QSBI's corporate headquarters, manufacturing, and/or assembly plants.

Corporate Headquarters in Raleigh, North Carolina. The head-quarters location has a headcount of 50 people, among which are the president of the corporation, and a corporate staff with responsibilities for accounting, legal counsel, environmental operations, health and safety, worldwide distribution, and worldwide purchasing. All of the corporate functions also exist at each location level; however, one person may actually perform several functions. The corporate staff focuses on overall company policies and leveraging of business opportunities.

QSBI headquarters are located in downtown Raleigh in an eight-story building. The staff occupies 25 percent of one floor (approximately 10,000 square feet). Attached to this building is a public parking structure, and the building is accessible through mass transit.

Metal Components Plant Outside McAllen, Texas. This site has a headcount of 300 people. Primary manufacturing and assembly activities at the QSBI McAllen site include:

- Design and test of products
- Receiving (receive metal rolls)
- Receiving (molded plastic parts from Mexico City and nylon straps from suppliers)
- Metal press (form metal seat belt and fastener components)
- Degreasing (remove oil, grease, and debris from metal components)
- Chrome electroplating (apply chrome surface to the metal components)
- Component and product packaging and shipping
- Assembly of components and seat belts
- Testing of components and seat belts

Facility operations and equipment include:

- Small industrial system for pretreatment of acids and metals in wastewater from electroplating operations
- Cooling towers
- Boilers
- Treated process water
- Small chemical-distribution and waste-handling facility
- Groundwater corrective action
- Small cafeteria with grease traps

The plant, which is owned by QSBI, was built in 1970 on 10 acres. The facility underwent a major expansion in 1985. During this expansion, it was discovered that the 10,000-gallon underground storage tank, which had been used to store degreasing solvent, was leaking. This tank was replaced immediately with a tank that had secondary containment, and corrective action for soil and groundwater remediation was initiated.

Manufacturing and Assembly Plant in Mexico City.   This site has a headcount of 350 people. Primary manufacturing and assembly activities include:

- Receiving (receive metal components from McAllen plant and nylon straps from suppliers)
- Receiving plastic beads
- Injection molding (form plastic seat belt components and fasteners)
- Component and product packaging and shipping
- Assembly of components and seat belts

Facilities operations and equipment include:

- Boilers
- Heating coils
- Sanitary discharge (no industrial wastewater discharge)
- Materials-distribution and waste-management facility

QSBI leases space in a suburb of Mexico City to manufacture molded plastic parts and assemble seat belts. The facility was originally a manufacturing location for textiles and has since been purchased and portioned off for lease by various manufacturing operations. QSBI rents 30 percent of the manufacturing space at this location.

Assembly Plants in São Paulo, Brazil, and Northern Milan, Italy. The São Paulo plant has a headcount of 200 people and the Milan plant has a headcount of 150 people. Both reside in leased space in a light industrial park. Primary activities at both locations include:

- Receiving from McAllen and Mexico City (fasteners and plastic parts)
- Receiving from suppliers (belts and specialty items)
- Assembly of seat belts
- Product packaging and shipping

QSBI's Two-Tiered Approach to Environmental Management

QSBI has a two-tiered EMS; that is, there is a corporate system that interacts with the individual systems at the plant sites. The definition of what is associated with each level of environmental management is shown in Fig. 2-3. As shown in the figure, corporate defines the "what" and the plant sites define the "how." This two-tiered approach will be carried throughout examples and illustrations for implementing an EMS, as applicable. In addition, QSBI has developed a total quality environmental management plan (TQEMP), which is presented in Fig. 2-4. This plan is referred to, as appropriate, in various chapters of the book.

Gap Analysis

QSBI performed an analysis of how its existing system aligns with the ISO 14001 standard and has identified some gaps. Where possible, the corporate environmental operations officer would like to fill as many gaps as possible with solutions that can be used by all sites within QSBI, in order to simplify the implementation process. The gaps identified by

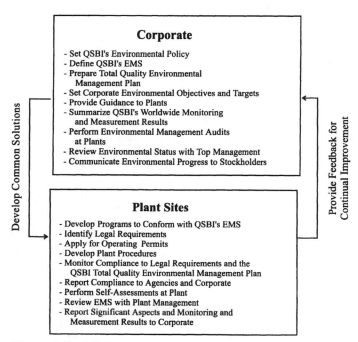

Figure 2-3. QSBI's two-tiered management system.

Quality Seat Belts, Inc.
Total Quality Environmental Management Plan

Introduction:

Quality Seat Belts, Inc. (QSBI) is committed to protection of the environment, conservation of natural resources, and pollution prevention. This plan has been developed to address specific areas of QSBI's operations that have potential to impact the environment and to provide corporate goals in these areas.

Environmental Goals:

(1) QSBI initiated a recycling program for paper and aluminum cans in 1994 at the McAllen, Texas, facility. Since that time, the recycling program for paper and aluminum cans has been expanded and now includes recycling of at least a portion of these commodities at all QSBI locations worldwide. QSBI's goal is to recycle 100% of these commodities by the year 2000.

(2) QSBI's operations consume energy, and the corporation is committed to reducing energy at its locations worldwide. QSBI's goal is to reduce energy consumption by 5% in the year 1999, using 1996 as a baseline.

Responsibilities:

QSBI's Environmental Operations Officer is responsible for coordinating the corporation's overall efforts to achieve this plan. The Location Environmental Coordinators are responsible for coordinating efforts at the location to achieve this plan.

Prepared by John Smith, Environmental Operations Officer   May 17, 1996

Figure 2-4. QSBI's total quality environmental management plan.

QSBI and the methods of closing these gaps—through either corporatewide or site-specific solutions—are delineated within each chapter. The last chapter of this book provides a summary of lessons learned through the process of integrating ISO 14001 into an existing EMS. The appendixes include generic ISO employee-training materials and a generic EMS audit methodology.

3

Section 4.2: Environmental Policy

A well-structured environmental management system (EMS) is predi-cated upon a strong environmental policy that is set by top manage-ment. Anyone who has worked in the business world knows how important top-management commitment is. Without it, the best-laid plans are often doomed for failure. The crafters of the ISO 14001 stan-dard recognized this and made top management responsible for defin-ing the organization's environmental policy. Indeed, many ISO 14001 experts think that this is the most prescriptive element of the standard. There are certain requirements that the environmental policy must meet, including the following:

- It must be appropriate to the nature, scale, and environmental impacts of the organization's activities, products, and services.
- It must include a commitment to continual improvement and pre-vention of pollution.
- It must include a commitment to comply with relevant environmen-tal legislation and regulations and with other (voluntary) require-ments to which the organization subscribes.
- It must provide a framework for setting and reviewing environmen-tal objectives and targets.
- It must be documented, implemented, and maintained.

- It must be communicated to all employees.
- It must be available to the public.

Policy Elements

Top-Management Commitment

Although ISO 14001 requires top management to define the organization's environmental policy, it does not provide a definition of "top management." As was discussed in Chap. 2, the standard applies to numerous types of organizations (see Fig. 2-1 in Chap. 2), and top management should be reflective of the organization implementing it. For example, if a multilocation corporation is implementing ISO 14001, top management will likely be the president, the chief executive officer, or the vice-president of environmental operations. If the organization is a single facility, then top management will likely be the plant manager. If the organization is a business unit within a multibusiness plant site, then top management would likely be the director of that business unit. Top management may also be defined as a group of managers, such as president, chief financial officer, and/or vice-president of environmental programs. However top management is defined, that person or persons must have the authority to commit money and resources to achieve policy objectives.

Appropriate to Nature and Scale of Activities, Products, and Services

The ISO 14001 standard allows an organization the flexibility to customize its environmental policy to fit its own needs. Although some policy statements are going to be common across all organizations—such as statements about complying with regulatory requirements, and commitments to continual improvement and prevention of pollution—it is expected that organizations with different activities, products, and services will have different policy objectives. Objectives that might be considered when setting or revising an environmental policy are presented in Fig. 3-1.

Commitment to Continual Improvement and Prevention of Pollution

The commitment to "continual improvement" is related to continual improvement of the EMS and not of environmental performance, per se, as was discussed in Chap. 2. In addition, the process of continual improvement does not have to take place in all areas of activities simultaneously. This allows an organization to prioritize which areas it will spend money and resources improving.

Objectives to Consider When Setting or Revising an Environmental Policy

✓ Commitment to minimize use of raw materials

✓ Commitment to minimize releases to air, water, and land

✓ Commitment to comply with all applicable laws and regulations

✓ Commitment to reuse, reduce, and recycle

✓ Commitment to use recycled products and renewable energy sources, where feasible

✓ Commitment to product stewardship

✓ Commitment to safeguard the environment for future generations

✓ Commitment to sustainable development

✓ Commitment to take corrective action, when appropriate

✓ Commitment to be a responsible neighbor

✓ Commitment to foster openness and dialogue with employees and the public on environmental issues and concerns

✓ Commitment to continual improvement of the environmental management system and/or environmental performance

✓ Commitment to perform self-assessments of compliance and/or environmental performance

---

Figure 3-1. Objectives that an organization might consider when setting or revising an environmental policy.

The term "prevention of pollution" is defined as "use of processes, practices, materials or products that avoid, reduce or control pollution, which may include recycling treatment, process changes, control mechanisms, efficient use of resources and material substitution" (ISO 14001, Section 3.6). It is fair to say that commitment to continual improvement of the EMS is linked to a commitment to prevention of pollution since improving processes, practices, and control mechanisms will aid in prevention of pollution. The concept of continual improvement of the EMS, however, goes beyond just improving the organization's infrastructure. It includes improving the overall management system, which could also include improvements in communication, training, the environmental policy, documentation, and other EMS elements. Methods of demonstrating commitment to continual improvement and to prevention of pollution are presented in Fig. 3-2.

**Methods to Demonstrate Commitment to
Continual Improvement and Prevention of Pollution**

Commitment to Continual Improvement

✓ Improved communications of environmental policy to employees and contractors

✓ Improved process for identifying environmental aspects and in setting objectives and targets

✓ The development of new training programs

✓ The development of operational procedures

✓ Tracking of additional key characteristics

✓ Improved preventive maintenance and equipment calibration programs

✓ Additional testing of emergency response procedures

✓ Redefined process for investigation and handling of nonconformance

✓ Improved environmental performance

✓ Improved compliance with legislation/regulatory requirements

✓ Improved EMS audit program

✓ Formalized management review process

Commitment to Prevention of Pollution

✓ Procedure for reviewing chemicals before purchase to ensure the least toxic chemical and appropriate volumes are bought

✓ Program for recycling of aluminum cans, paper, and plastics

✓ Program for investigating product end-of-life alternatives

✓ Reduction of chemical waste

✓ Program for using recycled materials in products

✓ Technology upgrades that maximize abatement

✓ Technology upgrades that provide for safer handling of hazardous materials

✓ Carpooling programs and flexible schedules for trip reductions

Figure 3-2. Examples of methods that an organization might use to demonstrate commitment to continual improvement and prevention of pollution.

Commitment to Comply with Relevant Environmental Legislation and Regulations and with Other Requirements

The organization's environmental policy must include a commitment to comply with relevant environmental legislation and regulations and with other requirements to which the organization subscribes. These "other requirements" might include voluntary programs such as the International Chamber of Commerce (ICC) Charter, U.S. Environmental Protection Agency's Energy Star program, Germany's Blue Angel program, industry standards, and the organization's own internal requirements. Essentially, under the ISO 14001 standard, if an organization subscribes to other requirements, these requirements are to be taken as seriously as legal and regulatory requirements.

Framework for Setting and Reviewing Environmental Objectives and Targets

The environmental policy needs to be specific enough that it provides a framework for setting and reviewing environmental objectives and targets. In other words, the policy forms the basis upon which other elements of the EMS should build, one element being the setting of objectives and targets. For example, if the organization generates a large amount of chemical waste, the policy should have a statement about reducing waste in order for it to be appropriate to the nature and scale of its activities. Based on this policy statement, the organization might want to set an objective and target with respect to waste. The objective might be to continually reduce waste generation over the next 5 years, and the target might be a reduction of waste of 3 percent per year, after adjusting for variation in production volumes.

Documented, Implemented, Maintained, and Communicated

Once the environmental policy is documented, top management must ensure that it is implemented throughout the organization, maintained, and communicated to all employees. The larger the organization, the bigger challenge this becomes, particularly the element of communication. Examples of how the environmental policy might be communicated in a large organization are presented in Fig. 3-3.

**Examples of Methods that Might Be Used
to Communicate the Environmental Policy
to Employees and Contractors**

✓ Review policy at new employee/contractor orientation

✓ Review policy during employee or contractor meetings

✓ Display policy on posters, bulletin boards, and/or tent cards in the cafeteria

✓ Provide information about policy in employee newsletters

✓ Include policy in work contracts

✓ Post policy near paper and/or aluminum can recycling areas

✓ Provide information about the policy during Earth Day activities

✓ Post policy in heavily trafficked areas, such as in the cafeteria or in the copier/FAX area

---

Figure 3-3. Examples of methods to communicate the organization's environmental policy.

Available to the Public

Lastly, but importantly, an organization's environmental policy must be available to the public. The public has become increasingly interested in environmental issues over the last several decades, and sharing the policy with the public is one way for top management to openly decree that protection of the environment is important. The policy can be made available to the public in several ways. Examples include placing the policy in stockholder reports, company promotional materials, local libraries, and on a worldwide web home page.

Environmental Policy of Quality Seat Belts, Inc.

Quality Seat Belts, Inc. (QSBI) instituted an environmental policy in 1980. The environmental policy was defined by QSBI's president (now retired) and has never been revised. Policy objectives included the following:

■ Protect the environment for future generations.

■ Apply practices and control technologies to minimize pollution.

■ Comply with current regulations worldwide.

■ Protect waterways from industrial pollution.

QSBI's Gap Analysis of Environmental Policy		
Requirement	*QSBI's Current Status*	*Comments*
Top management commitment	Conforms	Policy is defined by President.
Appropriate to nature and scale of activities, products and services	Improvements needed	Corrective action is not addressed; policy should emphasize recycling and reuse.
Commitment to continual improvement	Does not conform	Statement about continual improvement needs to be added.
Commitment to prevention of pollution	Conforms	Consider stronger statements with respect to prevention of pollution.
Commitment to comply to relevant legislation and other requirements	Improvement needed	QSBI internal requirements should be addressed.
Framework for objectives and targets	Improvement needed	The policy should address QSBI's present environmental situation.
Documented	Conforms	The policy was written in 1980.
Implemented	Assessment needed	Corporate should provide to locations a policy implementation assessment guidance.
Maintained	Does not conform	Policy should be reviewed with QSBI President.
Communicated	Does not conform	Plan for communication campaign needs to be established.
Available to public	Improvement needed	Consider publishing revised policy in QSBI external newsletter.

Prepared by John Smith, Environmental Operations Officer   February 5, 1997

Figure 3-4. QSBI's gap analysis for Section 4.2 of ISO 14001.

The corporate environmental operations officer of QSBI performed a gap analysis on the environmental policy, which is presented in Fig. 3-4. After completing the gap analysis, the corporate environmental operations officer met with the president of QSBI to review the environmental policy and the requirements of ISO 14001 with respect to the policy. As a result of the review, the policy was updated to include several new elements:

- Take corrective action when past practices have harmed the environment.
- Strive for continual improvement of the environmental management system.

Environmental Policy

Quality Seat Belts, Inc. (QSBI) is committed to protection of the environment at all locations where manufacturing and assembly activities are performed. As an environmentally conscious business, QSBI is committed to the following corporate environmental policy objectives:

- Fulfill the responsibility of trustee of the environment for this and future generations.

- To the extent possible, apply practices and control technologies that minimize pollution.

- Comply with current regulations worldwide. Set internal standards and requirements, as needed.

- Strive to minimize releases to air, water, and land.

- As appropriate, take corrective action where past practices have harmed the environment.

- Strive for continual improvement of the environmental management system.

- To the extent possible, use recycled materials throughout the company.

- Prevent pollution through reuse, recycling, and reduction.

Issued by James J. Wilson, President, QSBI   March 14, 1997

Figure 3-5. QSBI's environmental policy.

- To the extent possible, use recycled materials throughout the company.
- Prevent pollution through reuse, recycling, and reduction.

In addition, two elements were revised. The element about complying with current regulations worldwide was revised to include a commitment to set internal standards, as needed. The element about protecting waterways from industrial pollution was revised to address minimizing releases to air, water, and land.

QSBI's updated environmental policy is presented in Fig. 3-5. In order

Implementing the Environmental Policy at QSBI: Company-Wide Guidance

Fulfill the Responsibility of Trustee of the Environment for Future Generations

Applies to all QSBI locations, including manufacturing facilities, assembly plants, and Corporate Headquarters. Evidence of implementation includes activities such as sponsorship and/or participation in area environmental efforts; participation in local, regional and/or country environmental protection committees; and conformance to ISO 14001.

Apply Practices and Control Technologies that Minimize Pollution

Applies mainly to the manufacturing facilities and to a lesser extent to assembly plants. Evidence of implementation includes operational procedures and practices that minimize chemical usage and waste generation; use of proven wastewater treatment technologies; and use of safe chemical handling practices and procedures.

Comply with Current Regulations; Set Internal Standards Where Needed

Applies mainly to the manufacturing facilities. Evidence of implementation includes current licenses and permits for operation of processes; demonstration of compliance through successful internal or regulatory audits; and internal procedures that specify more stringent standards than those imposed by a particular government.

Minimize Releases to Air, Water, and Land

Applies mainly to manufacturing facilities and to a limited extent to assembly plants where releases to landfills occurred during refuse disposal. Evidence of implementation includes procedures/plans for release reduction; and records of release reductions through monitoring and measuring.

Continual Improvement of the EMS

Applies to all QSBI locations, including Corporate Headquarters, manufacturing plants, and assembly plants. Evidence of implementation includes improvement to any element of the EMS; improvement in technologies; and reductions in releases to air, water, and land.

Figure 3-6. Guidance for implementing elements of the environmental policy at QSBI.

Implementing the Environmental Policy at QSBI *Cont.*

Use Recycled Materials

Applies to all QSBI locations, including Corporate Headquarters, manufacturing plants, and assembly plants. Evidence of implementation includes use of recycled paper throughout the company; and use of recycled materials in seat belt and fastener products.

Prevent Pollution through Reuse, Recycling, and Reduction

Applies mainly to manufacturing facilities and to a lesser extent to assembly plants and Corporate Headquarters. Evidence of implementation includes aluminum can, paper, and plastics recycling programs through the company; water reuse programs; chemical reduction programs; and material reduction programs.

NOTE: The environmental policy is the foundation upon which all other elements of the environmental management system are built. For this reason, QSBI should have evidence at the Corporate and/or the Plant Site level that each policy element is being achieved.

Prepared by John Smith, Environmental Operations Officer   March 20, 1997

Figure 3-6. (*Continued*) Guidance for implementing elements of the environmental policy at QSBI.

to maintain the policy, the president will review it at least annually or whenever there is a significant change to QSBI's business process(es). (Note that the annual review is a commitment which would be auditable during the registration audit process.)

The environmental policy is issued at the corporate level, and all QSBI locations are expected to implement it. However, activities at the plant sites vary, so all policy elements may not actually be applicable to each site. Guidance for locations about ways to implement the various policy elements was developed by the corporate environmental operations officer, and is shown in Fig. 3-6.

PART 2
Planning

4

Section 4.3.1: Environmental Aspects

"Fail to plan and plan to fail" is an appropriate adage to keep in mind when setting up an environmental management system (EMS) that conforms to ISO 14001. After defining the environmental policy, planning is the next important step in the ISO 14001 process.

The planning process begins with the identification of environmental aspects and, subsequently, significant environmental aspects. This concept of identifying environmental aspects is unique to the ISO 14001 standard, and although many organizations may have considered environmental outcomes when setting priorities, few likely have developed procedures for a formalized process, as required by the standard.

The term "environmental aspect" is defined as an "element of an organization's activities, products, or services that can interact with the environment." There is a note associated with the definition that defines a "significant environmental aspect" as one "that has or can have a significant environmental impact" (Section 3.3). An "environmental impact" is defined as "any change to the environment, whether adverse or beneficial, wholly or partially resulting from an organization's activities, products, or services" (Section 3.4). Having provided definitions for environmental aspect and impact, the standard specifies that "the organization shall establish and maintain a procedure to identify environmental aspects of its activities, products or services that it can control or over which it can be expected to have an influence in order to determine those which have or can have significant impacts on the environment" (Section 4.3.1). This requirement brings the reader back to the note associated with "environmental aspects" which, in essence, calls aspects with significant environmental impacts, "significant aspects." It is important to note that the standard does not provide a definition for the term "significant." Thus the organization is able to use a variety of criteria to determine what is significant. Criteria for significance will be explored in the next section of this chapter.

The definitions and requirements related to environmental aspects are somewhat confusing, and it is not uncommon for this element to cause a substantial amount of "churn" when the organization begins implementing ISO 14001. The "churn" consists of lengthy discussions about the definition of an aspect; the definition of significant versus nonsignificant aspects and impacts; the need for a rating system; the number of people to include in the process; and other topics of conversation that encourage dissent, confusion, and chaos. Thus, in order to "stop the churn" and expedite the process, the authors have provided some guidance about how to develop the process of identifying environmental aspects and significant environmental aspects, which is detailed below.

Identifying Environmental Aspects

Beginning the Process

Although Annex A of ISO 14001 is for informative reference only (that is, its contents are guidance and not requirements), there is some useful information with respect to environmental aspects within it. Specifically, the annex suggests that "organizations should determine what their environmental aspects are, taking into account the inputs and outputs associated with their current and relevant past activities, products and/or services" (Section A.3.1).

This gives organizations a starting point, but churn can begin almost at once, if the persons identifying the aspects and significant aspects cannot agree to what detail the inputs and outputs should be categorized. Again, Annex A gives guidance by saying, "The process is intended to identify significant environmental aspects associated with activities, products or services, and is not intended to require a detailed life cycle assessment. Organizations do not have to evaluate each product, component or raw material input. They may select categories of activities, products or services to identify those aspects most likely to have a significant impact." Thus, by beginning with broad-based aspects associated with categories of activities, products, and services, the process can be simplified and still be effective. Examples of types of inputs and outputs that the organization might want to consider include those from normal operating conditions, shutdown and startup conditions, and foreseeable (realistic) emergency situations.

In addition to the discussion of how detailed the analysis for identifying significant aspects should be, another concern which often surfaces is what is the difference between an activity and an aspect. Examples of aspects that might fit both categories include chemical use, waste recycling, and transportation. At this early date in the ISO 14001 process, it is hard to get agreement about what constitutes an aspect versus an activity from those professionals who make it their business to know the explicit details of the standard; thus the authors have concluded that there is no "right" or "wrong" way to define it. Suffice it to say that if it is important to the EMS, it should be focused upon, either in this section of the standard or in operational control (see Chap. 13). Recommendations for managing this initial phase of the process are presented in Table 4-1.

Table 4-1. Recommendations for Managing the Process of Identifying Environmental Aspects

Do...

- Develop a process that is straightforward and repeatable.
- Make this process into a procedure.
- Ensure the process covers inputs and outputs of activities, products, and services.
- Start with broad-based aspects associated with categories of activities, products, and services and build detail from there.
- Consider beneficial as well as adverse environmental impacts.
- View this as a "first step," not a "final product." There is always an opportunity to enhance the list of environmental aspects at a later time.

Don't...

- Make the process into a life-cycle assessment of each product, component, or raw material input.
- Get bogged down debating the meaning of the definitions of environmental aspects and environmental impacts.
- Substantially change the process unless it is clearly not working.
- Make the process so complicated that it can't be performed in a reasonable amount of time. This is only one of many elements in the ISO 14001 standard, and all the other elements build from this, so don't get caught in the churn!

To further clarify the concepts described above, the following example can be used as guidance. An organization might identify energy use as an environmental aspect of its manufacturing activities. The amount of energy that the organization uses is something that it can control and have influence over. The organization purchases its energy from the only source available—a public utility—and the utility uses high-sulfur coal in its process. The organization does *not* have control over this. Thus the use of high-sulfur coal would not be an environmental aspect of the organization. However, if the organization uses a painting contractor to paint interior and exterior walls on its premises, the organization can be expected to have control and influence over this activity. The organization can control the paint type used (i.e., water-based versus solvent-based), spill-protection measures against potential releases, and waste generation and handling. For this activity, the organization could identify use of paints, potential for unplanned releases, and waste as environmental aspects.

Example environmental aspects (both adverse and beneficial) that might be identified by a manufacturing organization are presented in Fig. 4-1. Once environmental aspects are identified, the organization must determine which it considers significant environmental aspects.

Examples of Environmental Aspects Associated with Manufacturing Activities

✓ Planned releases to air, water, and/or land

✓ Unplanned releases to air, water, and/or land

✓ Contamination of land

✓ Consumption of raw materials

✓ Consumption of natural resources

✓ Generation of waste

✓ Emission of heat

✓ Creation of wildlife habitat*

   * Environmental aspects include those with beneficial as well as adverse impacts.

Figure 4-1. Examples of environmental aspects that might be identified by a manufacturing organization.

Identifying Significant Environmental Aspects

As mentioned earlier in this chapter, the term "significant" is not defined in the standard, and thus the organization has some latitude when developing criteria for determining which environmental aspects are significant environmental aspects. Criteria that might be used are presented in Fig. 4-2. Although not required, some organizations may want to put weighting factors on the criteria and perform a ranking of environmental aspects to determine which are significant. Another means of determining significance is through the judgment of environmental professionals. However the determination is made, environmental aspects that are identified as significant must be considered by the organization when it sets its objectives and targets. The following section details the process used at our hypothetical company, Quality Seat Belts, Inc. (QSBI), for identifying significant environmental aspects.

Criteria that Might Be Used by an Organization to Determine Significant Environmental Aspects

✓ Has a significant environmental impact based on frequency and/or severity

✓ Is covered under legislation or regulation

✓ Is covered by the organization's internal requirements

✓ Has potential to harm human health and/or the environment

✓ Is a concern to customers and/or the community

✓ Has a detrimental or beneficial effect on the natural beauty of the landscape

✓ Has the potential to affect climate

✓ Causes the depletion of natural resources

✓ Is covered by the organization's environmental policy

Figure 4-2. Criteria that might be considered when determining significant environmental aspects.

Environmental Aspects at QSBI

Although QSBI's total quality environmental management plan (TQEMP) addresses pollution prevention and sets forth some company-wide programs, it does not specifically address environmental aspects as defined in ISO 14001. The gap analysis, shown in Fig. 4-3, indicates that QSBI needs to establish and maintain a procedure for the identification of significant environmental aspects. In order to ensure consistency throughout QSBI, the environmental operations officer developed a procedure for identifying environmental aspects, including significant aspects, company-wide. This procedure is presented in Fig. 4-4.

QSBI's Gap Analysis of Environmental Aspects

Requirement	*QSBI's Current Status*	*Comments*
Establish and maintain a procedure to identify environmental aspects of activities, products, and services	Does not conform	Procedure is needed; establish one company-wide procedure.
Determine significant environmental aspects	Does not conform	Implement procedure and develop list of significant environmental aspects.
Ensure significant environmental aspects are considered when setting objectives and targets	Does not conform	Review Total Quality Environmental Management Plan (which delineates objectives and targets).
Keep the information up-to-date	Does not conform	Review list of significant environmental aspects annually.

Prepared by John Smith, Environmental Operations Officer   February 5, 1997

Figure 4-3. QSBI's gap analysis for Section 4.3.1 of ISO 14001.

Procedure Name: Identifying Significant Environmental Aspects
Document Control Number: QSBI Corporate Procedure 4.3.1
Document Owner: John Smith, Environmental Operations Officer

Introduction.

As part of the comprehensive environmental management system (EMS), QSBI has developed this procedure for the identification of environmental aspects of its activities, products, and services. The procedure outlines the requirements and responsibilities for identifying environmental aspects, including those which QSBI can control and those which QSBI can be expected to have an influence over, in order to determine which can have significant environmental impacts. Based on the information generated, QSBI will identify and document significant environmental aspects, which are considered when setting objectives and targets.

Requirements and Responsibilities.

The Environmental Operations Officer initiates the process of identifying environmental aspects for QSBI's activities, products, and services, and develops and documents a list of significant environmental aspects -- some of which are company-wide and some of which are specific to a plant site(s). The process includes soliciting input from Environmental Coordinators and professionals from other organizations (e.g., manufacturing, facilities, procurement, and distribution), as appropriate, and examining the design, manufacturing, assembly, and facilities activities at each location, including Corporate Headquarters.
    The following is considered:

a) inputs and outputs from routine operations;
b) inputs and outputs from major maintenance activities;
c) potential for accidents and emergency situations and their effect on the environment; and
d) inputs and outputs from on-premise services that could significantly affect the evironment.

The determination of significant environmental impacts will be based on the best professional judgment of the Environmental Operations Officer and the Environmental Coordinators, with advice and counsel from other professionals, as needed.

Figure 4-4. QSBI's procedure for an integrated approach to identifying significant environmental aspects.

Procedure Name: Identifying Significant Environmental Aspects *Cont.*
Document Control Number: QSBI Corporate Procedure 4.3.1

The determination will consider, at a minimum:

a) the environmental impact of the aspect;
b) legal/regulatory requirements of the country(ies) which have design, manufacturing, and assembly activities; and
c) QSBI's Total Quality Environmental Management Plan (TQEMP).

Environmental aspects which have or can have significant environmental impacts are classified as significant environmental aspects. Significant environmental aspects are documented, reviewed, and updated, as necessary. At a minimum, significant environmental aspects are reviewed annually.

The Environmental Operations Officer communicates the list of significant environmental aspects for all of QSBI to the Plant Site Environmental Coordinators. The Environmental Coordinators communicate the significant aspects that pertain to their location to all relevant persons. Any changes to the list of significant environmental aspects are likewise communicated.

References.

QSBI's TQEMP.

Records.

Records of identified significant environmental aspects.

Document History.

Date Issued: April 4, 1997.
Last Revision: Not Applicable.

Figure 4-4. (*Continued*) QSBI's procedure for an integrated approach to identifying significant environmental aspects.

Using the procedure, the team of environmental professionals identified activities at each site and then considered inputs and outputs (per the procedure) for each activity. A list of environmental aspects at each QSBI site—including corporate headquarters—was developed, and this list is presented in Table 4-2. Based on these environmental aspects, the team used the criteria listed in the procedure to develop a list of significant environmental aspects. Documentation of the use of the criteria is

Table 4-2. QSBI's Activities and Environmental Aspects

Corporate Headquarters

Activities: Administrative Office Activity Aspects:

- Energy use/consumption (building heating, cooling, and lighting)
- Material consumption (paper, office products)
- Nonhazardous waste (scrap paper, empty cartons and containers)
- Aluminum can and paper recycling

McAllen, Texas Plant

Activities: General Office Activity Aspects:

- Energy use/consumption
- Material consumption (paper, office products)
- Nonhazardous waste (scrap paper, empty cartons and containers)
- Aluminum can and paper recycling/waste avoidance

Activities: Electroplating Operations Aspects:

- Chemical use/consumption (plating chemicals)
- Water use/consumption
- Energy use/consumption
- Hazardous waste (sludge cake)
- Wastewater discharges
- Air emissions
- Unplanned releases

Activities: Metal Shop Operations Aspects:

- Chemical use/consumption (oils and degreasing agents)
- Energy use/consumption
- Material consumption (metal rolls)
- Metals recycling/waste avoidance
- Oil recycling/waste avoidance
- Degreasing solvent reuse/waste avoidance
- Nonhazardous waste (used oil and terpene-based solvents)

Activities: Assembly Operations Aspects:

- Materials consumption (metal, plastic, and nylon parts)
- Nonhazardous waste (plastic pieces, rags, and gloves)

Activities: Test Operations Aspects:

- Energy use/consumption

Activities: Shipping/Receiving/Packaging Aspects:

- Material consumption (boxes, pallets, shrink wrap)
- Recycling/waste avoidance (wood, cardboard, plastics)
- Nonhazardous waste (packaging scrap)

Table 4-2. QSBI's Activities and Environmental Aspects (*Continued*)

Activities: Facility Operations Aspects:

- Chemical use/consumption (water and wastewater-treatment chemicals, lawn-maintenance chemicals)
- Water use/consumption
- Energy use/consumption (boilers, chillers, building heating, cooling, and lighting)
- Nonhazardous waste (cafeteria grease and used oil)
- Wastewater discharges (cooling-tower blowdown)

Activities: Corrective Action Aspects:

- Groundwater remediation

Mexico City Plant

Activities: General Office Activity Aspects:

- Material consumption (paper, office products)
- Nonhazardous waste (scrap paper, empty cartons and containers)
- Aluminum can and paper recycling/waste avoidance

Activities: Plastic Shop Operations Aspects:

- Material consumption (plastic beads)
- Energy use/consumption
- Heat
- Nonhazardous waste (scrap plastic)

Activities: Assembly Operations Aspects:

- Material consumption (plastic, metal, and nylon parts)
- Nonhazardous waste (plastic pieces, rags, and gloves)

Activities: Shipping/Receiving/Packaging Aspects:

- Material consumption (boxes, pallets, shrink wrap)
- Recycling/waste avoidance (wood, cardboard, plastics)
- Nonhazardous waste

Activities: Facility Operations Aspects:

- Chemical use/consumption (water-treatment chemicals)
- Energy use/consumption (boilers; heating coils; building heating, cooling, and lighting)
- Nonhazardous waste (empty, rinsed containers)

Assembly Plants

Activities: General Office Activity Aspects:

- Material consumption (paper, office products)

Table 4-2. QSBI's Activities and Environmental Aspects (*Continued*)

- Nonhazardous waste (scrap paper, empty cartons and containers)
- Aluminum can and paper recycling/waste avoidance

Activities: Assembly Operations Aspects:

- Material consumption (plastic, nylon parts)
- Nonhazardous waste (plastic pieces, rags, and gloves)

Activities: Shipping/Receiving/Packaging Aspects:

- Material consumption (boxes, pallets, shrink wrap)
- Recycling/waste avoidance (wood, cardboard, plastics)
- Nonhazardous waste

Activities: Facility Operations Aspects:

- Energy use/consumption (building heating, cooling, and lighting)

Table 4-3. Assessment of Environmental Aspects for Classification as Significant

Aspect: Energy Use/Consumption

Environmental impact of aspects: Use of natural resources; impacts on air; impacts on global warming; potential impacts on cultural artifacts (from acid rain)

Legal/regulatory requirements: None

Addressed in TQEMP: Yes

Affected locations/plant sites: All

Significant corporatewide: Yes

Aspect: Material Consumption

Environmental impact of aspect: Use of natural resources

Legal/regulatory requirements: None

Addressed in TQEMP: No

Affected locations/plant sites: All for office activities; all (except corporate headquarters) for assembly activities; McAllen for metals; Mexico City for plastics

Significant corporatewide: No

Comments: Since this aspect is varied in its content (i.e., metals, plastics, boxes, office supplies), this aspect was assessed at the plant-site level

Significant at plant site: Yes—Mexico City (plastics)

Table 4-3. Assessment of Environmental Aspects for
Classification as Significant (*Continued*)

Aspect: Nonhazardous Waste (General)

Environmental impact of aspect: Impacts to land via landfill use

Legal/regulatory requirements: Not directly regulated, but limited landfill space in
Milan

Addressed in TQEMP: No

Affected locations/plant sites: All from minor office trash; all (except corporate
headquarters) from assembly scrap; McAllen for used oils and terpene-based solvents; Mexico City for plastic scrap

Significant corporatewide: No

Comments: Since this aspect is varied in its content (i.e., office trash, metal and
nylon scrap, used oils/solvent, and plastic scrap), this aspect was assessed at the
plant-site level

Significant at plant site: Yes—Mexico City (plastics)

Aspect: Aluminum Can and Paper Recycling/Waste Avoidance

Environmental impact of aspect: Reuse of materials; reduction of waste

Legal/regulatory requirements: No

Addressed in TQEMP: Yes

Affected locations/plant sites: All

Significant: Yes

Aspect: Recycling/Waste Avoidance (Other than Aluminum Cans and Paper)

Environmental impact of aspect: Reuse of materials; reduction of waste

Legal/regulatory requirements: No

Addressed in TQEMP: No

Affected locations/plant sites: All plant sites which have shipping/receiving activities; McAllen for metals and oil/solvent recycling

Significant corporatewide: No

Comments: Amounts minimal; assessed as significant on plant-site basis

Significant at plant site: No

Aspect: Chemical Use/Consumption

Environmental impact of aspect: Use of resources; potential for unplanned releases
that could impact the environment

Legal/regulatory requirements: Yes, as hazardous substances

Table 4-3. Assessment of Environmental Aspects for Classification as Significant (*Continued*)

Addressed in TQEMP: No

Affected locations/plant sites: McAllen and Mexico City

Significant corporatewide: No

Comments: Assessed as significant at plant sites based on usage amount and toxicity of chemicals

Significant at plant site: Yes—McAllen (includes all chemical use at plant site)

Aspect: Water Use/Consumption

Environmental impact of aspect: Depletion of natural resources

Legal/regulatory requirements: Restrictions in Mexico City

Addressed in TQEMP: No

Affected locations/plant sites: Specific to McAllen

Significant corporatewide: No

Comments: Assessed for significance by McAllen plant site

Significant at plant site: No

Comments: Water is not a focus item in this area; wastewater discharges have a higher priority (see Aspect: Wastewater Discharges)

Aspect: Wastewater Discharges

Environmental impact of aspect: Impacts to water quality

Legal/regulatory requirements: Yes

Addressed in TQEMP: No

Affected locations/plant sites: Specific to McAllen

Significant corporatewide: No

Comments: This aspect was assessed for significance by the McAllen plant site

Significant at plant site: Yes—McAllen

Aspect: Heat

Environmental impact of aspect: Temperature change in immediate area

Legal/regulatory requirements: No

Addressed in TQEMP: No

Affected locations/plant sites: Specific to Mexico City

Significant corporatewide: No

Table 4-3. Assessment of Environmental Aspects for
Classification as Significant (*Continued*)

Comments: Assessed for significance by the Mexico City plant site

Significant at plant site: No

Comments: Isolated operation under operational control

Aspect: Hazardous Waste

Environmental impact of aspect: Impact on land/air

Legal/regulatory requirements: Yes

Addressed in TQEMP: No

Affected locations/plant sites: Specific to McAllen (sludge cake)

Significant corporatewide: No

Comments: Assessed for significance by McAllen plant site

Significant at plant site: No

Comments: Amounts of hazardous waste generated are not significant

Aspect: Air Emissions

Environmental impact of aspect: Impact on air quality

Legal/regulatory requirements: Yes

Addressed in TQEMP: No

Affected locations/plant sites: Specific to McAllen

Significant corporatewide: No

Comments: Assessed for significance by McAllen plant site

Significant at plant site: Yes—McAllen

Comments: McAllen is in nonattainment area for air quality

Aspect: Groundwater Remediation

Environmental impact of aspect: Impacts (improvements) to groundwater

Legal/regulatory requirements: Yes

Addressed in TQEMP: No

Affected locations/plant sites: Specific to McAllen

Significant corporatewide: No

Comments: Assessed for significance by McAllen plant site

Significant at plant site: Yes—McAllen

Table 4-3. Assessment of Environmental Aspects for Classification as Significant (*Continued*)

Aspect: Unplanned Releases

Environmental impact of aspect: Impacts to air, water, and land

Legal/regulatory requirements: Yes

Addressed in TQEMP: No

Affected locations/plant sites: McAllen and (to a lesser extent) Mexico City

Significant corporatewide: No

Comments: Assessed for significance at McAllen and Mexico City plant sites

Significant at plant site: Yes—McAllen

Table 4-4. List of QSBI's Significant Environmental Aspects

Corporate Level (Corporate Headquarters and All Plant Sites)

- Aluminum can and paper recycling/waste avoidance
- Energy use/consumption

Specific to Plant Site

McAllen, Texas Plant Site

- Energy use/consumption
- Chemical use/consumption
- Wastewater discharges
- Air emissions
- Aluminum can and paper recycling/waste avoidance
- Groundwater remediation
- Unplanned releases

Mexico City Plant Site

- Material consumption
- Nonhazardous waste (plastic scrap)
- Aluminum can and paper recycling/waste avoidance
- Energy use/consumption

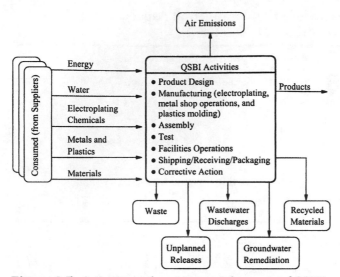

Figure 4-5. Activities and environmental aspects of QSBI's business.

presented in Table 4-3. A list of significant environmental aspects, as they pertain to the entire company and the specific plant sites, was documented, and this list is presented in Table 4-4. Finally, the team summarized their process in a flowchart that is presented in Fig. 4-5.

5
Section 4.3.2: Legal and Other Requirements

This element of ISO 14001 is succinct and to the point. It requires an organization to establish and maintain a procedure for identifying and providing access to legal requirements applicable to its activities, products, or services. Further, if the organization subscribes to other requirements—such as voluntary agency or industry requirements or internal company mandates—then these requirements must also be part of the procedure.

Examples of Legal Requirements

Although the rigor and depth of legal requirements pertaining to environmental management varies from country to country, most developed countries have mature regulatory schemes. Developing countries have less mature regulatory schemes, but they have the advantage of being able to review existing regulations for incorporation into theirs. Unfortunately, developing countries are sometimes too aggressive and put in place regulations that are not possible to meet because the infrastructure to support the regulation does not exist. For example, hazardous waste regulations may require the use of treatment technologies that are not yet available in the country.

Thus the standard emphasizes the *commitment* to comply with existing regulations in the environmental policy and does not make conformance to the standard dependent on *actual* compliance. Certainly, if an organization does not comply with all existing regulations, there should be plans in place for achieving compliance as soon as possible. In many ways, ISO 14001 will encourage environmental protection even if there is minimal regulatory agency support, since many developing countries will begin to use the standard as a vehicle to force companies to set and adhere to sound and common environmental practices in all countries with which the company conducts business.

Examples of environmental issues addressed by regulations in developed countries are presented in Table 5-1. As shown in the table, environmental regulations have evolved beyond traditional environmental

Table 5-1. Examples of Environmental Issues Addressed by Regulations in Developed Countries

Protection of the Environment

- Air
- Waterways and oceans
- Land and landscape

Protection of Flora and Fauna

- Forests
- Wilderness
- Fisheries
- Endangered species
- Marine mammals
- Wetlands

Environmental Management Issues

- Toxic substances
- Hazardous waste
- Pollution prevention
- Energy

Communication and Planning Issues

- Hazard information
- Emergency planning
- Environmental liability

issues such as protection of the environment, flora, and fauna, and now include issues such as environmental management, communication, and planning.

QSBI's Procedure to Identify and Have Access to Legal and Other Requirements

The gap analysis prepared by the environmental operations officer at QSBI for this element of the standard is presented in Fig. 5-1. As shown in the figure, QSBI locations had a process in place for identifying and providing access to legal requirements, but the process was not formalized in a procedure. Because the process varied slightly at the various locations, the environmental operations officer asked the plant site environmental coordinators to document the process used at their locations. Once documented, the procedure becomes part of the core elements of the location's environmental management system (EMS). The procedure developed by the environmental coordinator at the McAllen, Texas plant is shown in Fig. 5-2.

Corporate legal counsel provides guidance on environmental regulations impacting QSBI on an as needed basis. This person also leads the corporate audit team, which provides dual auditing functions of evaluating compliance with relevant environmental legislation and regulations and auditing the EMS at all QSBI plant sites. These separate audit functions are presented in Chaps. 15 and 18, respectively.

QSBI's Gap Analysis of Legal and Other Requirements

Requirement	*QSBI's Current Status*	*Comments*
Establish and maintain a procedure to identify and have access to legal and other requirements	Does not conform	Process is in place, but not documented in a procedure. Process varies slightly at different plant sites; thus, sites should document their processes separately.

Prepared by John Smith, Environmental Operations Officer   Feb. 10, 1997

Figure 5-1. QSBI's gap analysis for Section 4.3.2 of ISO 14001.

Procedure Name: Identifying and Providing Access to Legal and Other Requirements
Document Control Number: McAllen Plant Procedure 4.3.2
Document Owner: Javier Ramirez, QSBI Environmental Coordinator

Introduction.

A facility must comply with all legal and other requirements that apply to the site or its operations. Therefore, knowledgeable personnel must continuously monitor new requirements published by local, state and federal authorities to determine applicability for the site. Environmental programs and objectives have to be modified to incorporate new compliance requirements. Affected personnel must be given access to the requirements, and trained on their implications and impacts. The new requirements and associated changes to affected EHS programs must be added to the internal audit process to assure they have been implemented appropriately.

Requirements and Responsibilities:

Legal Requirements. The Environmental Coordinator receives a monthly update of regulatory changes from a reputable subscription service. Also he or she scans various websites to access new or proposed regulatory issues that may be of interest. The Environmental Coordinator is responsible for reviewing new and proposed regulations, making applicability determinations, and printing hard copies of regulations that apply to the site. The new requirements are placed in a binder to provide access to others, and be readily available to undergo further study so that modifications to environmental programs can be implemented to assure compliance. In addition, the Environmental Coordinator forwards new or proposed legal requirements to affected organizations (e.g., manufacturing, facilities, and distribution), for their review. These affected organizations are required to respond to the Environmental Coordinator within two weeks of their review as to the applicability of the regulations to their operations. The Environmental Coordinator and the affected organization assess the impact of the new requirements together and develop methods for meeting these requirements.

Regulatory requirements are typically included in work procedures, training materials, and emergency plans. The Corporate Audit Team is also notified so that new requirements are included in the site compliance audit process.

Other Requirements. The QSBI plant at McAllen, Texas is a member of Clean Texas 2000, a state voluntary program. The Environmental Coordinator is responsible for implementing the voluntary program requirements and for reviewing the program status annually.

Figure 5-2. McAllen's procedure for identifying and providing access to legal and other requirements.

Procedure Name: Identifying and Providing Access to Legal and Other Requirements *Cont.*
Document Control Number: McAllen Plant Procedure 4.3.2

References.

           Code of Federal Regulations (CFR).
           Texas Register.
           McAllen, Texas Local Codes and Ordinances.
           Websites:
                Federal Register - EPA Environmental Subset:
                    http://www.epa.gov/epahome/EPAFR-EnvSubset.html
                USEPA's Air Pollution Database (AIRS):
                    http://www.epa.gov/airs/airs.html
                National Institute of Standards and Technology (NIST):
                    htp://www.nist.gov/
                Environmental Resource Center (ERC):
                    http://ftp.clearlake.ibm.com/ERC/overview.html
                Envirolink:
                    http://envirolink.org/envirowebs.html

Records.

Application for Membership in Clean Texas 2000.
Clean Texas 2000 Membership Certificate.

Document History.

Document Issued: March 11, 1997.
Last Revision: Not Applicable.

Figure 5-2. (*Continued*) McAllen's procedure for identifying and providing access to legal and other requirements.

6

Section 4.3.3: Objectives and Targets

General Requirements

ISO 14001 requires an organization to establish and maintain documented objectives and targets. Elements to consider when establishing these objectives and targets include:

- Relevant legal requirements and other requirements to which the organization subscribes

- Significant environmental aspects of the organization's activities, products, and services

- Technological options available to the organization

- The organization's financial, operational, and business requirements

- The views of interested parties

This seemingly short list actually covers a broad range of topics, and takes into account legal and other requirements, the organization's specific business situation, and the views of interested parties.

This latter consideration—that of considering the views of interested parties—is worth some additional discussion. "Interested party" is defined in the standard as "individual or group concerned with or affected by the environmental performance of an organization" (Section 3.11). The term "concerned with" makes the definition very broad.

Examples of Interested Parties
✓ Neighborhood groups
✓ Environmental groups
✓ Citizen groups
✓ Political groups
✓ Governmental agencies
✓ Employees
✓ Stockholders
✓ Customers

Figure 6-1. Examples of interested parties.

Examples of who might be considered an interested party are presented in Fig. 6-1.

Although the standard does not directly require solicitation of the views of interested parties, it tacitly implies such by expecting the organization to have considered these views when setting its objectives and targets. Thus it would be desirable for the organization to establish some means for understanding the views of interested parties which are most appropriate (and palatable) for that organization. Examples of how this might be achieved are shown in Fig. 6-2.

Last, in addition to considering the elements above, the standard requires that the objectives and targets be consistent with the organization's environmental policy and with the commitment to prevention of pollution. As discussed in Chap. 1, the environmental policy is the first tier of the environmental management system (EMS) framework, and thus it is appropriate to relate objectives and targets to the policy. The commitment to prevention of pollution—also a key part of the EMS—is reemphasized here and in other elements of the standard.

Our hypothetical company, Quality Seat Belts, Inc. (QSBI), has developed a set of objectives and targets using the considerations defined above. QSBI's process used for arriving at their objectives and targets is discussed in the next section.

QSBI's Process for Setting Objectives and Targets

As with the other elements of the standard, QSBI's environmental operations officer performed a gap analysis to assess whether the company's

---

**Examples of Ways to Determine
Views of Interested Parties**

✓ Attendance by organization representative at neighborhood meetings, town board meetings, or other local meetings

✓ Participation by organization representative on local emergency planning committee

✓ Sponsorship of facility open house

✓ Availability of hot line for community input to the facility

✓ Providing tours of the facility to governmental agency representatives, environmental organization leaders, and other persons

✓ Participation in community/business activities, such as information exchanges, Earth Day activities, and household waste management activities

✓ Customer surveys

Figure 6-2. Examples of ways an organization can determine views of interested parties.

present EMS was adequate to conform to this section of the standard. This gap analysis is shown in Fig. 6-3. As shown in the figure, QSBI has set two goals in its total quality environmental management plan (TQEMP) (see Chap. 2, Fig. 2-4) which could constitute objectives and targets. These include:

1. Recycle 100 percent of paper and aluminum cans at all QSBI locations worldwide by the year 2000.

2. Develop programs for reducing energy use at QSBI's locations worldwide by 1999.

This gives QSBI a good start, but there is still some work to be done in this area. First, the wording of the goals could better reflect specific objectives and targets as defined in the standard. The standard defines "environmental objective" as "overall environmental goal, arising from the environmental policy, that an organization sets itself to achieve, and which is quantified where practicable" (Section 3.7). An "environmental target" is defined as "detailed performance requirement, quantified where practicable, applicable to the organization or parts thereof, that arises from the environmental objectives and that needs to be set and

QSBI's Gap Analysis of Objectives and Targets

Requirement	QSBI's Current Status	Comments
Documented objectives and targets are established and maintained	Conceptually conforms	Environmental goals established in the TQEMP do not specifically reflect the ISO 14001 standard.
The following were considered when setting objectives and targets:		
legal and other requirements	Conforms	The goals address hazardous waste, nonhazardous waste, and recycling, per regulations and Clean Texas 2000 goals.
significant environmental aspects	Does not conform	Significant environmental aspects were not defined when the TQEMP was developed.
technological options	Conforms	Technology availability was considered when setting goals.
financial, operational, and business requirements	Conforms	These were considered when setting goals.
views of interested parties	Conforms	The views of the general public toward voluntary pollution prevention were considered. More specific views of the local community need to be addressed.
Objectives and targets are consistent with environmental policy	Needs improvement	Objectives and targets need to be updated with respect to revised environmental policy.
Objectives and targets are consistent with the commitment to prevention of pollution	Conforms	Prevention of pollution is a main element of the policy, objectives and targets.

Prepared by John Smith, Environmental Operations Officer.   Feb. 17, 1997

Figure 6-3. QSBI's gap analysis for Section 4.3.3 of ISO 14001.

met in order to achieve those objectives" (Section 3.10). When looked at from this perspective, the two goals developed by QSBI could be turned into objectives and targets as described below.

Goal 1 could become:
Objective: Increase recycling of paper and aluminum cans at QSBI.

Targets:

1. Recycle 60 percent of all paper and aluminum cans at QSBI in 1997.
2. Recycle 75 percent of all paper and aluminum cans at QSBI in 1998.
3. Recycle 90 percent of all paper and aluminum cans at QSBI in 1999.
4. Recycle 100 percent of all paper and aluminum cans at QSBI in 2000.

Goal 2 could become:
Objective: Reduce energy usage at QSBI location worldwide.

Targets:

1. Identify areas of energy consumption in 1997.
2. Initiate energy-conservation programs in 1998.
3. Reduce energy consumption by 5 percent in 1999.

A second gap identified in Fig. 6-3 is that QSBI did not consider its significant environmental aspects when setting objectives and targets. QSBI has recently identified significant environmental aspects on a location basis (see Chap. 4), so these need to be reviewed for possible inclusion in this part of the EMS. (Note: Objectives and targets do not have to be set for each significant environmental aspect, but the standard requires the organization to consider them for inclusion.) Finally, the environmental policy has recently been revised and expanded (see Chap. 3), and the new elements should also be considered for inclusion in QSBI's objectives and targets. Thus QSBI decided to update its environmental objectives and targets by listing company-wide and location-specific objectives and targets, and placing them in a document entitled QSBI Environmental Management Program (described in the next chapter). This document, once developed, will supersede the TQEMP. The current list of QSBI's environmental objectives and targets is presented in Table 6-1.

Table 6-1. QSBI's Environmental Objectives and Targets

QSBI Worldwide Objectives and Targets	
Objectives	Targets
Increase recycling of paper and aluminum cans	1. Recycle 60% in 1997 2. Recycle 75% in 1998 3. Recycle 90% in 1999 4. Recycle 100% in 2000
Minimize energy usage	1. Identify areas of energy consumption in 1997 2. Initiate energy-conservation programs in 1998 3. Reduce energy consumption by 5% in 1999

Location-Specific Objectives and Targets: McAllen, Texas Plant	
Objective	Targets
Reduce chemical usage—relative to production	Reduce chemical usage by 5% in 1997
Reduce air emissions—relative to production	Reduce air emissions by 3% in 1997

Location-Specific Objective and Targets: Mexico City Plant	
Objective	Targets
Reduce material consumption (plastics); reduce nonhazardous waste (plastics)—relative to production	1. Reduce plastic consumption/waste by 5% in 1997 2. Reduce plastic consumption/waste by 20% in 1998

7

Section 4.3.4: Environmental Management Program

Developing an Effective Environmental Management Program

Once the organization sets its environmental objectives and targets, the ISO 14001 standard requires the organization to establish and maintain an environmental management program to achieve them. (Note: the use of the term "environmental management program" in the standard relates singularly to a program for achieving objectives and targets and not to what is conventionally considered an environmental management program such as a waste management program, air monitoring program, and others.) The program must include "the who," "the when," and "the how." In other words, the following must be defined:

- The designation of responsibility for achieving objectives and targets at each relevant function and level of the organization.
- The time frame by which they are to be achieved.
- The means by which they are to be achieved.

The environmental management program must be updated as new developments or modifications to activities, products, and services warrant.

There are two confusing (and some might contend controversial) areas in this section of the standard. First, the phrase "each relevant function and level of the organization" is not specifically defined in the standard; thus it is left up to the organization to determine the appropriate function and/or level at which to assign responsibility for achieving objectives and targets. This must be carefully thought out because pushing the responsibility to the lowest level within the organization (i.e., the individual level) can cause inconsistencies, while keeping responsibility at too high a level within the organization (i.e., the management level or environmental staff level) can cause inaction.

The second confusing element in this section of the standard is the requirement to define the "mean(s)" for achieving objectives and targets. Traditional definitions of the term "means" include two that seem to be fitting (Funk and Wagnall's, 1968): (1) "the medium, method, or instrument by which some end is or may be accomplished"; and (2) "money, property, or other resources." These two definitions are actually markedly different in nature. Using the first definition, the standard would imply that the environmental management program would need to include some type of work plan for each relevant function and level; whereas, using the second definition, the standard would imply that it is sufficient to identify budget line items and/or resources in the environmental management program. The authors suggest that a combination of these terms be used, to the extent practical, so that the environmental management program is dependent on the depth and/or breadth of the objectives and targets and the relevant functions and levels involved. Consider the following scenario. An objective of the organization is the reduction of energy consumption, and the target is to reduce overall energy use at a facility by 3 percent within 12 months. This objective and target may affect the entire population of the organization, if individual actions of turning off lights and computers make a difference in bottom-line energy use. If this is the case, then, responsibility for achieving the objective and target would be at the individual level. Thus all employees should know about the energy-conservation objective and target and how they can conserve energy; however, it is impractical for every individual to have a work plan of how he or she will conserve energy. Instead, the energy conservation work plan could be developed at a location level, by the facilities coordinator or other appropriate person, and communicated to all employees. The work plan might consist of the following:

- Communication of the energy objective and target to all employees.

- Communication of what is expected of employees to conserve energy (e.g., turn off lights when not in the office and power off computers and printers at the end of the day).

- Investigate the use of low-energy lighting that meets building code for hallways and exit signs.

- Install automatic shutoffs on building lights during evening hours and weekends.

To make the environmental management program complete, the work plan would need to include the person responsible for each action and budget allocations (as appropriate), as well as the time frame.

On the other hand, the organization may set for itself an objective of 3 percent energy reduction over each of the next 3 years, as follows:

- Year 1—implement reduction measures in facility operations.

- Year 2—implement reduction measures in manufacturing operations.

- Year 3—implement reduction measures in building/office areas.

In this case, relevant functions will differ from year to year, as follows:

- Year 1—relevant functions include the facilities staff.

- Year 2—relevant functions include the facilities staff and the manufacturing engineers.

- Year 3—relevant functions include facilities staff, manufacturing engineers, and employees.

What is important here is to clearly define who is responsible for each element of the energy-reduction objective and target. In this example, the facilities staff would be expected to implement measures to reduce energy use by 3 percent in Year 1, and then maintain operational control of the reduction measures in Years 2 and 3. Manufacturing engineers would be expected to implement measures to reduce energy use by 3 percent in Year 2, and then maintain operation control of the reduction measures in Year 3. Finally, facilities staff and employees would likely be expected to implement measures for a reduction in energy by 3 percent in Year 3. Again, the work plan would need to include the person responsible for each action and budget allocations (as appropriate)—in this case the time frame is already defined.

In another scenario, the objective is to minimize unplanned releases and the target is to install secondary containment around outside tanks by the year 2000. In this case, the objective and target are managed by one department—the facilities operations department. This department develops the work plan and controls the budget, resources, and schedule to achieve the objective and target. Aside from the purchasing department (if the organization has a separate purchasing function), no other function or level needs to be included in developing and implementing the environmental management program for this objective and target.

In a final scenario, the objective is to minimize the volume of waste going to landfill and the target is to initiate a program for recycling empty chemical drums. Responsibilities for this include the following:

- The environmental coordinator is responsible for finding a vendor who will recycle the empty chemical drums and a transporter to take the drums from the facility to the vendor.

- The purchasing department is responsible for negotiating the contracts for the empty drum recycling and transport activities.

- The waste-handling technicians are responsible for triple rinsing the drums to make them ready for the recycling vendor.

Thus the relevant functions and levels associated with this objective and target include the environmental coordinator, the purchasing department, and the chemical services department. The operator of the plating department is responsible for placing the empty chemical drum in a designated area; however, this must be done regardless of whether the drum is recycled or disposed. Therefore, the responsibility is associated with general environmental management, and not this specific objective and target. However, the authors encourage communication whenever possible, so that personnel are aware of the objectives and targets, even when their actions are not expected to impact them directly.

Last, the organization should ensure that the objectives and targets are realistic. If they are not met during the time frame specified in the environmental management program, there should be a plan in place to ensure meeting them in the near future. If the organization determines that they are too aggressive, then new objectives and targets should be established.

Recommendations for implementing this important element of the standard are presented in Table 7-1. An example environmental management program prepared by the hypothetical company, Quality Seat Belts, Inc. (QSBI), is presented below.

Table 7-1. Recommendations for Implementing the Environmental Management Program

Do...

- Include "the who," "the when," and "the how" for achieving each objective and target.
- Document the program.
- Clearly and carefully determine the relevant function and level for achieving the objectives and targets.
- Carefully consider whether a work plan is needed and, if so, at what level it should be developed.
- Keep the program practical.
- Communicate the program to all persons involved, even those on the periphery.
- Amend the program in light of new developments in activities, products, and services.

Don't...

- Keep the environmental management program too high or push it too low within the organization.
- Put "stretch goals" or impractical projects in the program—be realistic.
- Neglect the program—keep it current.
- Forget that continual improvement is key to the standard—everything doesn't have to be done at once.

QSBI's Environmental Management Program

QSBI performed a gap analysis of its total quality environmental management plan (TQEMP) (see Chap. 2, Fig. 2-4) versus what was required by ISO 14001 for the environmental management program, and this gap analysis is presented in Fig. 7-1. As shown in the figure, QSBI will need to update the document to include new objectives and targets, and better definitions of relevant levels and functions, as well as the means for achieving all objectives and targets. The updated document (now called QSBI's environmental management program) is presented in Fig. 7-2.

QSBI's Gap Analysis of Environmental Management Program
(Currently Termed Total Quality Environmental Management Plan)

Requirement	*QSBI's Current Status*	*Comments*
Program for achieving objectives and targets includes:		TQEMP does not address all objectives and targets
designation of responsibility at each relevant function and level	Needs improvement	Each relevant function and level is not adequately addressed.
the time frame	Conforms	A more detailed time frame of when the environmental management program is to be revised is needed.
the means	Does not conform	The means for achieving objectives and targets is not included.
Program is amended to include new or modified activities, products, or services	Not applicable	The environmental management program will be amended to include new objectives and targets, as needed, during continual improvement activities.

Prepared by John Smith, Environmental Operations Officer   February 17, 1997

Figure 7-1. QSBI's gap analysis for Section 4.3.4 of ISO 14001.

QSBI'S Environmental Management Program
Document Control Number: QSBI Corporate Document 4.3.4
Document Owner: John Smith, Environmental Operations Officer

QSBI's Corporate Objectives and Targets

Objective	Target	Person Responsible	Time Frame	Means
Increase recycling of paper and aluminum cans	(1) recycle 60%	Environmental Operations Officer	1Q 1997	Set up contracts with recycling vendors.
		Plant Site Environmental Coordinator	1Q 1997	Install paper and aluminum can recycling bins -- cost $3,500.
		Plant Site Environmental Coordinator	1Q 1997	Communicate to employees specifics of the program.
		Employees	2Q 1997	Begin following program.
	(2) recycle 75%	Plant Site Environmental Coordinator	1Q 1998	Initiate recycling awareness campaign; assess need for additional recycling bins.
		Environmental Operations Officer	2Q 1998	Evaluate progress of program for corporation; determine the need for additional activities.
		Employees	During year, 1998	Follow program.

QSBI's Environmental Management Program *(Cont.)*
Document Control Number: QSBI Corporate Document 4.3.4

QSBI'S Corporate Objectives and Targets (Cont.)

Objective	Target	Person Responsible	Time Frame	Means
Increase recycling of paper and aluminum cans (cont.)	(3) recycle 90%	Plant Site Environmental Coordinator	1Q 1999	Initiate recycling campaign.
		Environmental Operations Officer	1Q 1999	Alter cleaning contract to include segregation of office waste.
		Cleaning Contractors	2Q 1999	Begin office waste segregation activities.
		Employees	During year, 1999	Follow program.
Minimize energy usage	(1) identify areas of energy consumption	Plant Site Facilities Coordinator	2Q 1997	Install additional energy meters -- cost $5,000.
	(2) initiate energy conservation program	Environmental Operations Officer/Plant Site Facilities Coordinator	2Q 1998	Develop program.
		Plant Site Department Supervisors	3Q 1998	Communicate program to employees

Figure 7-2. QSBI's environmental management program.

QSBI's Environmental Management Program *(Cont.)*
Document Control Number: QSBI Corporate Document 4.3.4

Corporate Objectives and Targets (Cont).

Objective	Target	Person Responsible	Time Frame	Means
Minimize energy usage (cont.)	(3) reduce energy consumption by 5%	Employees	During year, 1999	Operational control.

Objectives and Targets Specific to McAllen, Texas Plant

Objective	Target	Person Responsible	Time Frame	Means
Reduce chemical usage -- relative to production	reduce by 5%	Environmental Coordinator	2Q 1997	Review chemical usage at plant.
		Facilities Coordinator	3Q 1997	Install flow meters on high volume baths -- cost $1,000.
		Department Supervisors	3Q 1997	Reissue process specs, as needed.
Reduce air emissions -- relative to production	reduce by 3%	Environmental Coordinator	2Q 1997	Review efficiency of scrubber.
		Facilities Coordinator	3Q 1997	Optimize scrubber efficiency.

QSBI's Environmental Management Program *(Cont.)*
Document Control Number: QSBI Corporate Document 4.3.4

Objectives and Targets Specific to the Mexico City Plant

Objective	Target	Person Responsible	Time Frame	Means
Reduce material consumption (plastics)/ Reduce nonhaz-ardous waste (plastics) -- relative to production	(1) reduce by 5%	Environmental Coordinator	2Q 1997	Select and purchase equipment to reuse plastic scrap in plastics mold process -- cost $2,500.
		Plastics Shop Supervisor	3Q 1997	Install reuse equipment on plastics mold process; instruct employees on use.
		Employees	4Q 1997	Begin reusing plastic scrap in process.
	(2) reduce by 20%	Employees	During year, 1998	Operational control.

Change History:

Document Issued: May 8, 1997
Last Revision: Not Applicable

Figure 7-2. *(Continued)* QSBI's environmental management program.

PART 3

Implementation and Operation

8

Section 4.4.1: Structure and Responsibility

Thus far, the general requirements, environmental policy, and planning sections of the ISO 14001 standard have been reviewed. As was noted in Part 1, top-management commitment is of paramount importance if the environmental management system (EMS) is to succeed. Also of importance is the appointment of a representative(s) by top management for planning, implementing, operating, monitoring, and correcting the EMS. This representative(s) must also report on the performance of the EMS to top management for review and as a basis for improvement of the EMS.

Defining Roles and Responsibilities

Once again, the ISO 14001 standard is flexible in its approach when identifying requirements for structure and responsibility. Exactly who has the authority and/or responsibility for the various duties that facilitate an effective EMS is left up to management to define, based on existing circumstances of the organization. There is no one "right" way to define roles, responsibilities, and authorities. However, the formation of the definitions is needed to make the EMS system-dependent and not person-dependent, which is one of the many strengths inherent in ISO

Requirements Pertaining to Structure and Responsibility

✓ Roles, responsibilities, and authorities must be defined, documented, and communicated.

✓ Management must provide resources essential to the implementation and control of the EMS.

✓ Resources should include human resources and specialized skills, technology, and financial resources.

✓ Top management must appoint representative(s) to establish, implement, and maintain the EMS.

✓ Top management must appoint representative(s) to report on the performance of the EMS to them for review and as a basis for improvement of the EMS.

Figure 8-1. Requirements defined in ISO 14001 for structure and responsibility.

14001. A summary of the requirements of this section of the standard is presented in Fig. 8-1.

As shown in the figure, the first step to conforming with this section of the standard is to define, document, and communicate roles, responsibilities, and authorities. This latter requirement of communication is typical of many elements of ISO 14001 since communication at all relevant levels is what drives environmental responsibility down to the core of the organization. Examples of ways to communicate these roles, responsibilities, and authorities are presented in Fig. 8-2. The second requirement in this section of the standard is to provide resources necessary to implement and control the EMS, and these should include human, technological, and financial resources. Finally, top management must appoint a specific representative(s) who, irrespective of other duties, tends to the EMS and reports status and progress to top management. The following section details how the hypothetical company Quality Seat Belts, Inc. (QSBI) implements this part of the standard.

Implementation of Structure and Responsibility at QSBI

QSBI performed a gap analysis of structure and responsibility, which is shown in Fig. 8-3. As shown in the figure, QSBI initially had only the

---

**Examples of Ways to Communicate Roles,
Responsibilities, and Authorities**

✓ Department pictures on the bulletin board with names and responsibilities within the EMS

✓ Department or functional meetings wherein roles, responsibilities,
and authorities for the EMS are presented to every employee

✓ EMS awareness campaign developed and presented to top management, middle management, line management, environmental
professionals, and line employees

✓ On-line access to EMS organization and responsibilities charts

✓ Employee newsletters which provide information about the structure and responsibility of the EMS

Figure 8-2. Communicating roles, responsibilities, and authorities.

rudiments of structure and responsibility in place. As a result, QSBI
clearly defined and documented all major roles, responsibilities, and
authorities for the EMS at the corporate and location levels. The
updated responsibility matrix at the corporate level is presented in Fig.
8-4, and plant site-specific responsibility matrices are presented in Fig.
8-5.

QSBI's Gap Analysis of Structure and Responsibility

Requirement	QSBI's Current Status	Comments
With respect to the EMS, roles, responsibility, and authorities are:		
defined	Needs improvement	Environmental Operations Officer's and Plant Site Environmental Coordinators' roles are clearly defined; other responsibilities are not.
documented	Needs improvement	Organization charts are current but do not include all roles and responsibilities with respect to the EMS.
communicated	Needs improvement	Once revised, roles, responsibility, and authorities should be communicated throughout corporation.
Proper resources are provided	Conforms	Resources include human resources, technology, and financial resources.
Representatives are appointed and have defined roles, responsibility, and authorities for:		
ensuring EMS requirements are established, implemented, and maintained in accordance to ISO 14001	Conforms	Environmental Operations Officer and Plant Site Environmental Coordinators do this.
reporting on the performance of the EMS to top management	Needs improvement	Top management reviews the performance of the EMS, but not on a scheduled basis.

Prepared by John Smith, Environmental Operations Officer   February 21, 1997

Figure 8-3. QSBI's gap analysis for Section 4.4.1 of ISO 14001.

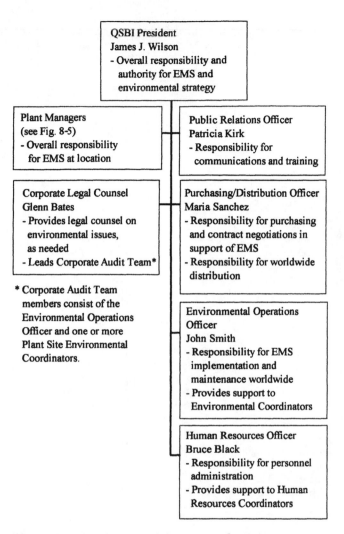

Figure 8-4. EMS responsibility matrix for QSBI at the corporate level.

Plant Manager McAllen, Texas	Plant Manager Mexico City	Plant Manager São Paulo	Plant Manager Milan
Environmental Coordinator - Implements EMS - Coordinates emergency response - Supervises chemical activities	Environmental Coordinator - Implements EMS - Coordinates emergency response - Supervises facilities and maintenance activities	Environmental Coordinator - Implements EMS - Coordinates emergency response - Supervises facilities and maintenance activities - Coordinates training - Coordinates communications	Environmental Coordinator - Implements EMS - Coordinates emergency response - Supervises facilities and maintenance activities - Coordinates training - Coordinates communications
Human Resources Coordinator - Coordinates communications - Coordinates training	Human Resources Coordinator - Coordinates communications - Coordinates training	Shipping/Receiving Coordinator - Coordinates receiving, shipping, and packaging activities - Provides purchasing services to support EMS	Shipping/Receiving Coordinator - Coordinates receiving, shipping, and packaging activities - Provides purchasing services to support EMS
Purchasing Coordinator - Provides purchasing services to support EMS	Purchasing Coordinator - Provides purchasing services to support EMS	Assembly Department Supervisor - Supports EMS at department level	Assembly Department Supervisor - Supports EMS at department level
Facilities Coordinator - Supervises facilities and maintenance activities - Repairs equipment	Department Supervisors for plastics shop, assembly, shipping and receiving - Support EMS at department level		
Department Supervisors for plating, assembly, metals shop, test, shipping and receiving - Support EMS at department level			

Figure 8-5. EMS responsibility matrices for QSBI at the plant-site level.

9

Section 4.4.2: Training, Awareness, and Competence

General Requirements

Training of relevant personnel is integral to proper functioning of the environmental management system (EMS). Since training is dependent on the organization's activities, products, and services, the organization must identify training needs. The ISO 14001 standard requires that all personnel whose work may impact the environment receive appropriate training. In addition, the organization must establish and maintain procedures in order to make employees or members at each relevant level aware of:

- The importance of conformance with the environmental policy and procedures and with the requirements of the EMS.

- The significant environmental impacts, actual or potential, or their work activities and the environmental benefits of improved personal performance.

- Their roles and responsibilities in achieving conformance with the environmental policy and procedures and with the requirements of the environmental management system, including emergency preparedness and response requirements.

- The potential consequences of departure from specified operating procedures.

Finally, the standard requires that personnel performing tasks which could cause a significant impact must be competent to perform that task, based on education, training, or experience. Examples of training and awareness needs that a typical manufacturing facility might put into its training plan are presented in Fig. 9-1. Example training overheads that might be used in an EMS general awareness training program are presented in Appendix A.

Training, Awareness, and Competence for Quality Seat Belts, Inc. (QSBI)

The hypothetical company, QSBI, performed a gap analysis for this element of the standard, which is presented in Fig. 9-2. As shown in the figure, several of the requirements are already met within existing programs. These include a training plan and records for employees whose work may create a significant environmental impact, emergency response training, and job descriptions which outline education, experience, and/or training needed to perform the job function.

From the gap analysis, QSBI determined that what was still needed was a formalized procedure for training and awareness that addresses all the requirements in the standard. This was developed by the environmental operations officer and is presented in Fig. 9-3. In addition, a training/awareness program which addresses the environmental policy, the significant environmental aspects, and other elements of the EMS was identified as being needed for employees throughout the corporation. Material to include in the training program was developed by the environmental operations officer in conjunction with the human resources officer, and is presented in Table 9-1.

Examples of Training and Awareness Needs That a Typical Manufacturing Facility Might Identify

Training Needs

✓ Training on hazardous and nonhazardous waste management procedures

✓ Training on emergency response procedures

✓ Training on proper handling of chemicals

✓ Training on operational procedures, such as wastewater treatment operations, maintenance, and facilities operations

✓ Training on monitoring and measuring methods

✓ Training on calibration methods and schedule for monitoring and measuring equipment

✓ Training on proper packaging of chemicals and waste for transport

✓ Training on proper completion of shipping papers

✓ Contractor training on the facility's environmental policy, procedures, and the EMS

✓ EMS auditor training

Awareness Needs

✓ Awareness about the environmental policy, significant environmental aspects of the facility's activities, products, and services, and the requirements of the EMS

✓ Awareness of the core documents of the EMS and methods for accessing these

✓ Awareness of how to access applicable legal requirements

✓ Awareness of pollution prevention efforts

✓ Awareness of how the employees' job responsibilities can impact the environment

✓ Awareness of how the employees' job responsibilities are related to the facility's objectives and targets, environmental management program, and continual improvement of the EMS

✓ Awareness of the environmental benefits of improved performance and the potential consequences of deviating from specified operating procedures

---

Figure 9-1. Examples of training and awareness to include in a typical training plan.

**QSBI's Gap Analysis of Conformance to
Training, Awareness, and Competence**

Requirement	*QSBI's Current Status*	*Comments*
Identify training needs	Needs improvement	Training is identified and coordinated at the Corporate and plant site levels; however, training with respect to the EMS needs to be identified, developed, and coordinated.
Personnel whose work may create a significant cnvironmental impact have received proper training	Conforms	Training plans and records exist.
Procedures are in place to make employees aware of:		
the importance of conformance with the environmental policy, procedures, and the EMS	Does not conform	Need procedures and awareness training program.
the environmental impacts of their work and the benefits of improved performance	Needs improvement	Employees are aware, but no formalized awareness training has been deployed.
their roles and respon-sibilities in relation to the environmental policy and the EMS	Does not conform	Awareness training needs to be deployed.

Figure 9-2. QSBI's gap analysis for Section 4.4.2 of ISO 14001.

**QSBI's Gap Analysis of Conformance to
Training, Awareness, and Competence *Cont.***

Requirement	QSBI's Current Status	Comments
Procedures are in place to make employees aware of:		
their role and responsibilities with respect to emergency response requirements	Conforms	All employees are trained in emergency evacuation; identified employees are trained in spill response.
potential consequences of departure from specified operating procedures	Needs improvement	Employees are aware of potential environmental consequences, but procedure is not in place.
Personnel who perform tasks that could cause significant environmental impact are competent in their job.	Conforms	Job descriptions and necessary education, training, or experience is available for each position.

Prepared by John Smith, Environmental Operations Officer   February 21, 1997

Figure 9-2. (*Continued*) QSBI's gap analysis for Section 4.4.2 of ISO 14001.

Procedure Name: Training, Awareness, and Competence
Document Control Number: QSBI Corporate Procedure 4.4.2
Document Owner: John Smith, Environmental Operations Manager

Introduction.

QSBI considers training, awareness, and competence paramount to ensuring a strong
environmental management system (EMS). Thus, there are requirements for training,
awareness and competence of all employees throughout the corporation whose job functions
could impact the environment and the EMS.

Requirements and Responsibilities for Training and Awareness.

Requirements and responsibilities for training, awareness, and competence reside at the
Corporate and plant site levels.

Corporate Headquarters Responsibilities. At the Corporate level, the Environmental
Operations Officer is responsible for developing the corporate wide environmental training
plan, which meets regulatory requirements and those of the EMS. Further, the Environmental
Operations Officer is responsible for developing a training module for EMS awareness
training. The Human Resources Officer is responsible for coordinating the training efforts
worldwide and for maintaining training records of the Corporate staff.

At the plant sites, responsibilities for training, awareness, and competence are as follows:

McAllen, Texas and Mexico City Plant Sites. At these plant sites, the Environmental
Coordinator reviews the Corporate Headquarters training, awareness, and competence plan
and adds location-specific requirements, as necessary, including all legally required training.
The Human Resources Coordinator is responsible for coordinating the training efforts and for
maintaining training records for the location.

Milan, Italy and São Paulo, Brazil Plant Sites. At these plant sites, the Environmental
Coordinator reviews the Corporate Headquarters training, awareness, and competence plan
and adds location-specific requirements, as necessary, including all legally-required training.
The Environmental Coordinator is responsible for coordinating the training efforts and for
maintaining training records for the plant sites.

Figure 9-3. QSBI's procedure for training, awareness, and competence.

Procedure Name: Training, Awareness, and Competence *Cont.*
Document Control Number: QSBI Corporate Procedure 4.4.2

Requirements and Responsibilities for Competence.

Job descriptions are written for all employees which describe education, experience, and training needed to perform that job function. In addition, training requirements are defined in the Corporate and location training plans. The Coordinator or Supervisor of the job is responsible for ensuring that employees have the proper education, experience, and training.

References.

None.

Records.

Corporate and location training plans.
Corporate training records.
Location training records.
Job descriptions.

Document History.

Date Issued: June 2, 1997.
Last Revision: Not Applicable.

Figure 9-3. (*Continued*) QSBI's procedure for training, awareness, and competence.

Table 9-1. Material to Include in the EMS Training Program for QSBI Employees

General Background

- What is ISO 14001?
- What are the key elements of the standard?
- What procedures are required by the standard?
- What are the benefits of implementing ISO 14001?
- What does QSBI's EMS look like?
- What is QSBI's plan for implementation of ISO 14001?
- What are employees' roles and responsibilities?
- Keys to success during implementation

Environmental Policy

- What is QSBI's environmental policy?
- How can employees and contractors support this policy?

Planning a Strong and Successful EMS

- QSBI's significant environmental aspects—what are they?
- Understanding legal and regulatory requirements
- QSBI's objectives and targets
- What part do employees and contractors have in achieving objectives and targets?
- Implementing an environmental management program

Implementing and Operating QSBI's EMS

- QSBI's organization, roles, responsibilities, and authority for the EMS
- Employee training, awareness, and competence requirements
- Communication within and outside of QSBI
- Environmental management system documentation—what are the core elements of QSBI's EMS?
- Controlling environmental documents
- Controlling operations—what are the key procedures?
- Accidents or incidents—what are QSBI's procedures for handling emergencies?

Ensuring QSBI's EMS Is Working—Checking and Corrective Action

- Monitoring and measurement systems at QSBI
- Handling any identified nonconformance—including putting in place corrective and preventive action
- Records retention
- Auditing QSBI's EMS

Table 9-1. Material to Include in the EMS Training Program for QSBI Employees (*Continued*)

Management Review
■ Is the system suitable, adequate, and effective?   ■ How can we continually improve?

10

Section 4.4.3: Communication

General Requirements

Communication—particularly internal communication—is one of the most important elements of the environmental management system (EMS). The ISO 14001 Standard requires organizations to establish and maintain procedures for both internal and external communication about the significant environmental aspects and the EMS. Internal communication is expected to be "multidirectional"—not just from the top down, but from the bottom up and throughout all relevant functions and levels of the organization. The organization must also have a documented process for receiving, documenting, and responding to relevant external communication. Further, the organization must consider processes for external communication about its significant aspects and record its decision about which, if any, it will deploy. Essentially, this requires the organization to consider processes for voluntary reporting to the public, although the organization is allowed to elect not to voluntarily report at all. Examples of methods for communicating about the EMS both internally and externally are presented in Fig. 10-1. An example procedure for internal and external communications is presented in the next section of this chapter, which describes how the hypothetical company Quality Seat Belts, Inc. (QSBI) conforms to this element of the standard.

**Methods for Communicating about
Environmental Aspects and the EMS**

Internal Communication Methods

✓ Department or functional meetings which review significant environmental aspects and progress toward achieving objectives and targets

✓ Employee newsletters which communicate elements of the EMS

✓ Internal Web page

✓ Periodic internal reports about the status of the EMS

✓ A publicized internal phone number to provide information about the EMS and/or to allow feedback or recommendations for improvements

✓ Area wall charts that depict environmental measurements

✓ Ongoing communications between the environmental staff and relevant functions/levels with respect to the environmental management program and efforts toward prevention of pollution and continual improvement

External Communication Methods

✓ External environmental performance reports

✓ Communication through stockholder reports

✓ An external communications hot line

✓ External Web page

✓ Presentations at industry and/or governmental meetings about the organization's EMS and environmental performance

Figure 10-1. Examples of internal and external communication methods.

Internal and External Communication at QSBI

The environmental operations officer at QSBI—the hypothetical company implementing ISO 14001—performed a gap analysis on this segment of the standard. This gap analysis is presented in Fig. 10-2. As shown in the figure, QSBI already has a procedure for receiving inquiries about environmental matters, although the procedure needed

QSBI's Gap Analysis for Internal and External Communications		
Requirement	*QSBI's Current Status*	*Comment*
With respect to environmental aspects and the EMS, procedures are established for:		
internal communication between the various levels and functions of QSBI	Partially conforms	Communications take place but are not proceduralized.
receiving, documenting, and responding to relevant communication from external interested parties	Conforms	Public Relations has established a procedure for handling external inquiries about environmental issues.
QSBI has considered processes for external communication about its significant environmental aspects	Does not conform	QSBI prepares and submits reports required by governmental agencies, but has not considered voluntary external communication methods.
QSBI has recorded its decision about external communication about its significant environmental aspects	Does not conform	QSBI has not recorded a decision about voluntary external communication methods.
Prepared by John Smith, Environmental Operations Officer		February 21, 1997

Figure 10-2. QSBI's gap analysis for Section 4.4.3 of ISO 14001.

to be put in the document control system. This procedure, with updated information pertaining to document control, is presented in Fig. 10-3.

QSBI's gap analysis also revealed that there are several requirements not in place that needed to be addressed in order for QSBI to be in conformance with the standard. For example, although QSBI has established communication methods about the EMS at relevant functions and levels within the corporation, these were not documented in a procedure. The environmental operations officer developed a procedure for internal communications, and asked each environmental coordinator to do the same for relevant functions at their sites. These procedures for

Procedure Name: Responding to External Inquiries About Environmental Matters

Document Control Number: QSBI Corporate Procedure 4.4.3A
Document Owner: John Smith, Environmental Operations Officer

Introduction.

Responding to external inquiries about environmental matters pertaining to QSBI is important to the corporation. In order to ensure that responses are accurate and consistent throughout the corporation, responsibilities for answering these external inquiries are defined at the corporate level and at the plant site levels.

Requirements and Responsibilities.

Corporate Headquarters. At the Corporate level, all external inquiries and/or information requests are directed to the Public Relations Officer. If the inquiry pertains to an environmental matter, the Public Relations Officer contacts the Environmental Operations Officer, and they jointly draft a response. Inquiries and/or information requests are received through letters or telephone calls and can be from customers, government agency representatives, media representatives, and others who may be interested in QSBI's environmental activities. In addition, a plant site Human Resources Coordinator or Environmental Coordinator, as appropriate, has the responsibility to forward to the Public Relations Officer any non routine requests, such as requests for environmental performance trends and non published data.

The Public Relations Officer maintains a log of external inquiries, and maintains records of responses to those which might affect QSBI's business or public image. The Public Relations Officer and/or the Environmental Operations Officer informs the President of these communications at least monthly.

A copy of any response that pertains to a specific plant site is sent to that site's Human Resources Coordinator or the Environmental Coordinator, as appropriate

McAllen, Texas and Mexico City Plant Sites. At these sites, all external inquiries and requests for information are directed to the Human Resources Coordinator. If the inquiry or request for information pertains to an environmental matter, the Human Resources Coordinator contacts the Environmental Coordinator. Any routine requests, such as requests for data submission by government agency representatives, are responded to by the Environmental Coordinator, with a copy submitted to the Human Resources Coordinator. Any non-routine information requests are referred to the Corporate Public Relations Officer for response, as appropriate. The Human Resources Coordinator maintains a log of external inquiries pertaining to environmental matters and responses, as appropriate.

Figure 10-3. QSBI's corporate procedure for responding to external inquiries about environmental matters.

Procedure Name: Responding to External Inquiries About Environmental Matters
Cont.
Document Control Number: QSBI Corporate Procedure 4.4.3A

Milan, Italy and São Paulo, Brazil Plant Sites. At these sites, all external inquiries and requests for information are directed to the Environmental Coordinator. If the inquiry or request for information pertains to an environmental matter, the Environmental Coordinator handles it, as follows. Any routine requests, such as requests for data submission by government agency representatives, are responded to. Any non-routine information requests are referred to the Corporate Public Relations Officer for response, as appropriate. The Environmental Coordinator maintains a log of external inquiries pertaining to environmental matters and responses, as appropriate.

References.

QSBI Corporate Procedure 5: Public Relations.

Records.

Corporate Headquarters log of external inquiries.
Plant sites logs of external inquiries.
Responses to communications.

Document History.

Date Issued: April 7, 1993.
Last Revision: September 10, 1997.
Changes: Added document control number and document owner.

Figure 10-3. (*Continued*) QSBI's corporate procedure for responding to external inquiries about environmental matters.

corporate and for the Mexico City plant are presented in Figs. 10-4 and 10-5, respectively.

Another gap was that, although QSBI prepares and submits all reports required by governmental agencies, it had not considered processes for voluntary external communication about its significant environmental aspects. QSBI's environmental operations officer met with the president and presented the following as potential methods for doing this:

Procedure Name: Internal Communication About Significant Environmental Aspects and the EMS
Document Control Number: QSBI Corporate Procedure 4.4.3B
Document Owner: John Smith, Environmental Operations Officer

Introduction.

Internal communication about QSBI's significant environmental aspects and the EMS is necessary in order to maintain a suitable, adequate, and effective EMS. Communications at the Corporate level are multi-directional: that is clear lines of communication exist from Corporate Headquarters to plant site, from the plant sites to Corporate Headquarters, and within Corporate Headquarters.

Requirements and Responsibilities.

The Environmental Operations Officer is responsible for establishing and maintaining clear lines of communication about QSBI's significant aspects and the EMS. This is established and maintained, as follows:

a. The Environmental Operations Officer is responsible for communicating information about the EMS to the plant Environmental Coordinators. This information includes, but is not limited to, communications about environmental policy updates, significant environmental aspects, the environmental management program, the EMS audit, nonconformances identified within the EMS, and the results of management reviews. This information is communicated through an annual internal EMS progress report, which is written by the Environmental Operations Officer.

b. The Environmental Operations Officer communicates to the President and the Public Relations Officer information about environmental activities or external inquiries that may affect QSBI's business or public image.

c. The Environmental Operations Officer communicates to the Corporate Legal Counsel any legal/regulatory issues associated with QSBI's activities.

d. The Environmental Operations Officer communicates training needs for establishing, implementing, and maintaining the EMS to the Public Relations Officer.

e. The Environmental Operations Officer maintains an e-mail address and an internal phone number for receiving questions, information, and other communications from the plant Environmental Coordinators and Corporate Staff.

Figure 10-4. QSBI's corporate procedure for internal communications.

Procedure Name: Internal Communication About Significant Environmental Aspects and the EMS *Cont.*
Document Control Number: QSBI Corporate Procedure 4.4.3B

References.

None.

Records.

Records of communications.
Distribution lists of Environmental Coordinators at the plant sites.

Document History.

Date Issued: April 8, 1997.
Last Revision: Not Applicable.

Figure 10-4. (*Continued*) QSBI's corporate procedure for internal communications.

Procedure Name: Internal Communication About Significant Environmental Aspects and the EMS
Document Control Number: QSBI Mexico City Plant Procedure 4.4.3
Document Owner: Guadalupe Martinez, QSBI Environmental Coordinator

Introduction.

Internal communication about QSBI's significant environmental aspects and the EMS is necessary in order to maintain a suitable, adequate, and effective EMS. Internal communications include communications to Corporate Headquarters and communications to relevant functions and levels within the plant site. These are defined in this procedure.

Requirements and Responsibilities.

The Environmental Coordinator is responsible for establishing and maintaining clear lines of communication within the plant site and to Corporate Headquarters. This is established and maintained, as follows:

a. The Environmental Coordinator is responsible for communicating information about QSBI's EMS to the relevant functions and levels at the plant. This includes communication about:

- the Corporate environmental policy
- signficiant aspects
- objectives and targets
- the environmental management program
- information presented in the annual QSBI EMS progress report
- the EMS audit

b. The Environmental Coordinator communicates to the Corporate Environmental Operations Officer, the Plant Manager, and/or the Public Relations Coordinator information about the plant site's environmental activities, legal/regulatory status, and/or topics of concern, as warranted.

c. The Environmental Coordinator communicates training needs for establishing, implementing, and maintaining the EMS to the Human Resources Coordinator.

d. The Environmental Coordinator maintains an e-mail address and an internal phone number for receiving questions, information, and other communications from coordinators, supervisors, plant personnel, and Corporate Headquarters.

Figure 10-5. Mexico City's procedure for internal communications.

Procedure Name: Internal Communication About Significant Environmental Aspects and the EMS *Cont.*
Document Control Number: QSBI Mexico City Plant Procedure 4.4.3

References.

QSBI Corporate Procedure 4.4.3B.

Records.

Records of communications.

Document History.

Date Issued: April 16, 1997.
Last Revision: Not Applicable.

Figure 10-5. (*Continued*) Mexico City's procedure for internal communications.

- An annual communication to the public in the form of an environmental report.
- Quarterly environmental newsletters that are sent to employees and the community.
- No voluntary external communications about significant environmental aspects.

The president decided that QSBI should publicize the revised environmental policy, but at this time, the corporation should not commit to voluntary communications about its significant environmental aspects. This decision was documented in a memorandum from the president to the environmental operations officer.

11

Section 4.4.4: Environmental Management System Documentation

General Requirements

The requirements for environmental management system (EMS) documentation look deceptively simple. In essence, the standard requires the organization to establish and maintain information that describes the core elements of the EMS and their interaction, and to provide direction to related documentation. The information can be in paper or electronic form.

This is actually a formidable task. In meeting these requirements, the organization must first define the core elements of the EMS, which are those documents that the organization relies upon to define the way it manages its entire environmental system. Certainly, the environmental policy would be considered a core element. Other documents that are part of the EMS, but which are not considered "core" must also be listed or otherwise referenced. Examples of what might constitute core elements and documents that interact with the core elements are presented in Fig. 11-1.

Although not specifically required by the standard, many organizations are choosing to develop an EMS manual. This manual can describe

**Examples of Core Elements and Documents
That Interact with These**

Documents That Might Be Considered Core Elements of the EMS

✓ Environmental policy

✓ EMS manual

✓ Corporate or top-level procedures or instructions

✓ Directives from top management defining key EMS requirements

Documents That Might Interact with the Core Elements

✓ Department operating procedures

✓ Master equipment calibration schedule

✓ Controlled inspection forms

✓ Controlled reporting forms

✓ Guidance manuals

Figure 11-1. Examples of what might constitute core elements of an EMS and documents that interact with these.

in detail how the EMS system works, and can introduce the documents that interact (i.e., supporting documents) with the core elements. In the ISO 9000 world—that of quality system management—management system documentation and document control requirements are very strictly defined. The drafters of ISO 14001 took special care not to be prescriptive with these requirements since they wanted the focus of the EMS standard to be on environmental management, not "paper management." Nonetheless, organizations that have mature EMS's might want to give consideration to fully documenting the system—beyond what is required in the standard—as it will give the various parts of the system consistency and effectiveness. A pictorial view of how ISO 14001 documents and records might interact, in line with the ISO 9000 documentation pyramid, is presented in Fig. 11-2.

The next section details how the hypothetical company—Quality Seat Belts, Inc. (QSBI)—defines its core elements of the EMS. It is worth repeating at this point that the standard is broad and allows for customization of an EMS to fit the specifics of the organization. Thus there is no "right" way to implement this or any other section of the standard.

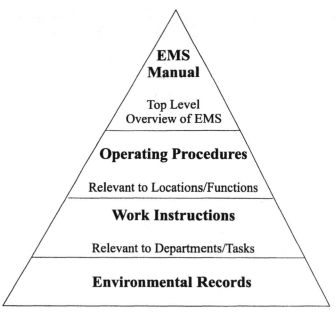

Figure 11-2. Environmental management system documentation pyramid.

Nevertheless, because of the ISO 9000 experience, if an organization decides to become registered to ISO 14001, there is typically added pressure to have this element meet predetermined expectations.

EMS Documentation at QSBI

Since the time QSBI began implementing ISO 14001, several documents have been updated and/or developed. Thus, when the environmental operations officer performed the gap analysis of this section of the standard—which is presented in Fig. 11-3—there was still work to be done, but much of it was in progress.

As shown in the gap analysis, the core elements of the EMS needed to be formally identified, and direction to related documentation needed to be provided by corporate and plant sites. A pictorial of the core elements and related documents and records was developed by the environmental operations officer to meet this element of the standard, and is presented in Fig. 11-4, including Attachments A, B, and C.

QSBI's Gap Analysis for Environmental Management System Documentation

Requirement	*QSBI's Current Status*	*Comments*
Establish and maintain information to:		
describe the core elements of the EMS and their interaction	Partially conforms	The core elements exist or are being developed. They have not been formally identified.
provide direction to related documentation	Does not conform	Direction to related documentation at Corporate and plant sites needs to be provided.

Prepared by John Smith, Environmental Operations Officer   February 28, 1997

Figure 11-3. QSBI's gap analysis for Section 4.4.4 of ISO 14001.

Page 1 of 4

QSBI's EMS Documentation Hierarchy
Document Control Number: QSBI Corporate Document 4.4.4
Document Owner: John Smith, Environmental Operations Officer

QSBI's EMS Documentation Hierarchy is defined in Attachments A, B, and C of this document.

Change History.

Date Issued: March 12, 1997.
Last Revision: Not Applicable

Figure 11-4. QSBI's EMS documentation hierarchy.

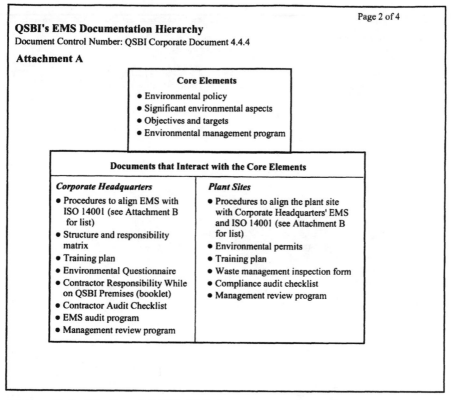

Figure 11-4. (*Continued*) QSBI's EMS documentation hierarchy.

QSBI's EMS Documentation Hierarchy
Document Control Number: QSBI Corporate Document 4.4.4

Attachment B -- List of Procedures/Programs at Corporate Headquarters and Plant Sites that Interact with QSBI's Core Elements of the EMS

Procedures at Corporate Headquarters

- QSBI Corporate Procedure 4.3.1 Identifying Significant Environmental Aspects
- QSBI Corporate Document 4.3.4 Environmental Management Program
- QSBI Corporate Procedure 4.4.2 Training, Awareness, and Competence
- QSBI Corporate Procedure 4.4.3A Responding to External Inquiries About Environmental Matters
- QSBI Corporate Procedure 4.4.3B Internal Communications About Significant Environmental Aspects and the EMS
- QSBI Corporate Procedure 4.4.5 Document Control
- QSBI Corporate Procedure 4.4.6 Procedure for Operational Control of Contractors
- QSBI Corporate Procedure 4.5.1 Monitoring and Measurement
- QSBI Corporate Procedure 4.5.2 Nonconformance and Corrective and Preventive Action
- QSBI Corporate Procedure 4.5.3 Records
- QSBI Corporate Procedure 4.5.4A EMS Audit
- QSBI Corporate Document 4.5.4B EMS Audit Program

Procedures at Plant Sites (as applicable)

- Plant Site Procedure 4.3.2 Identifying and Providing Access to Legal and Other Requirements
- Plant Site Procedure 4.4.3 Internal Communication About Significant Aspects and the EMS
- Plant Site Procedure 4.4.7 Emergency Preparedness and Response
- Plant Site Procedure 4.5.1A Monitoring and Measurement Plant Site Procedure 4.5.1B Periodically Evaluating Compliance with Environmental Legislation and Regulations
- Plant Site Procedure 4.5.1C Equipment Calibration

Figure 11-4. (*Continued*) QSBI's EMS documentation hierarchy.

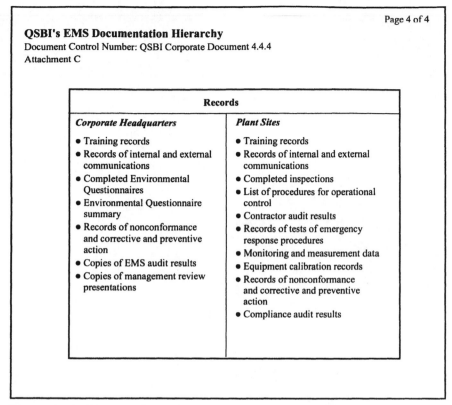

QSBI's EMS Documentation Hierarchy
Document Control Number: QSBI Corporate Document 4.4.4
Attachment C

Records	
Corporate Headquarters	*Plant Sites*
<ul><li>Training records</li><li>Records of internal and external communications</li><li>Completed Environmental Questionnaires</li><li>Environmental Questionnaire summary</li><li>Records of nonconformance and corrective and preventive action</li><li>Copies of EMS audit results</li><li>Copies of management review presentations</li></ul>	<ul><li>Training records</li><li>Records of internal and external communications</li><li>Completed inspections</li><li>List of procedures for operational control</li><li>Contractor audit results</li><li>Records of tests of emergency response procedures</li><li>Monitoring and measurement data</li><li>Equipment calibration records</li><li>Records of nonconformance and corrective and preventive action</li><li>Compliance audit results</li></ul>

Figure 11-4. (*Continued*) QSBI's EMS documentation hierarchy.

12
Section 4.4.5: Document Control

General Requirements

Under the ISO 14001 standard, documents are different from records, and thus are controlled differently. Documents are procedures, manuals, forms, and other documentation that is relied upon to demonstrate current or planned activities. Records, on the other hand, show evidence that an event or activity occurred. Examples include memoranda, completed forms, meeting minutes, and presentation materials. The topic of records is addressed in Chap. 17.

Once the core elements of the environmental management system (EMS) are identified, along with documents that interact with these, these documents must be controlled. The standard requires the organization to establish and maintain procedures to ensure the following:

- Documents can be located.

- Documents are periodically reviewed, revised as necessary, and approved for adequacy by authorized personnel.

- Current versions of the documents are available where needed.

- Obsolete documents are promptly removed or otherwise assured against unintended use.

- Obsolete documents retained for legal and/or knowledge-preservation purposes are suitably identified.

- Responsibilities are established concerning the creation and modification of the various types of documents.

Further, documents must be legible, dated, and readily identifiable, maintained in an orderly manner, and retained for a specified period.

Document Control at Quality Seat Belts, Inc. (QSBI)

The environmental operations officer at QSBI performed the gap analysis for this section of the standard, which is presented in Fig. 12-1. As shown in the figure, QSBI needed to put in place a procedure for controlling documents. This procedure was written by the environmental operations officer for the entire corporation, and is presented in Fig. 12-2.

QSBI's Gap Analysis of Document Control

Requirement	*QSBI's Current Status*	*Comments*
There is an established procedure for controlling documents that are core to the EMS and that interact with these. The procedure meets the requirements outlined in Section 4.4.5 of ISO 14001	Does not conform	QSBI has Corporate and plant site level documents that are not controlled per a formalized procedure.

Prepared by John Smith, Environmental Operations Officer   February 28, 1997

Figure 12-1. QSBI's gap analysis for Section 4.4.5 of ISO 14001.

Procedure Name: Document Control

Document Control Number: QSBI Corporate Procedure 4.4.5
Document Owner: John Smith, Environmental Operations Officer

Introduction.

Control of documents that are key to effective management of QSBI's environmental management system (EMS) is necessary to ensure that these documents are handled consistently throughout the corporation.

Requirements and Responsibilities.

Responsibilities for document control are as follows:

a. Core elements of the EMS and those documents that interact with these are listed in QSBI's EMS Documentation Hierarchy. The Corporate Environmental Operations Officer is responsible for establishing and maintaining this documentation hierarchy.

b. At Corporate Headquarters, the Environmental Operations Officer is responsible for creating and maintaining the procedures and documents defined in the EMS Documentation Hierarchy.

c. At the plant sites, the Environmental Coordinator is responsible for creating and maintaining the procedures and documents defined in the EMS Documentation Hierarchy.

d. The Environmental Operations Officer or the Environmental Coordinator, at their respective level, is responsible for ensuring that:

- all controlled documents can be located
- all controlled documents are reviewed at least annually, beginning April 1, 1997, and revised as necessary
- current versions of the documents are available where needed for effective functioning and management of the EMS
- obsolete documents are removed promptly from all points of use
- permits and other documents retained for legal/knowledge preservation are labeled "Obsolete: For Reference Only"

Figure 12-2. QSBI's procedure for document control.

Procedure Name: Document Control *Cont.*
Document Control Number: QSBI Corporate Procedure 4.4.5

- controlled documents are legible
- controlled documents are dated, with dates of revision
- maintained in an orderly manner
- retained for one year after they are revised

A master copy of all controlled Corporate documents is located in the office of the Environmental Operations Officer. A master copy of all controlled Plant Site documents is located in the office of the Environmental Coordinator.

References.

QSBI's EMS Hierarchy of Documentation.

Records.

Copies of controlled documents.

Change History.

Date Issued: May 7, 1997.
Last Revision: Not Applicable.

Figure 12-2. (*Continued*) QSBI's procedure for document control.

13

Section 4.4.6: Operational Control

General Requirements

Operational control involves all employees whose job functions have the potential to cause a significant impact on the environment. As such, these employees play a key role in the proper functioning of the environmental management system (EMS). Section 4.4.2 of ISO 14001 requires that employees be made aware of the environmental impacts of their actions and potential impacts from deviating from established procedures (see Chap. 9). This section, then, describes how the procedures for day-to-day operations are established and maintained.

The first requirement of this section is for the organization to identify operations and activities—including maintenance—associated with significant environmental aspects, and in line with the policy and objectives and targets. Once identified, the organization must then ensure that these operations and activities are carried out under specified conditions. In order to do that, the organization must:

- Establish and maintain documented procedures to cover situations where their absence could lead to deviations from the environmental policy and objectives and targets.
- Stipulate operating criteria in the procedures.

- Establish and maintain procedures related to identifiable significant aspects of goods and services used by the organization.

- Communicate relevant procedures and requirements to suppliers and contractors.

Documented operational control procedures and task instructions have long been part of most environmental management systems—whether they conform to other parts of ISO 14001 or not. What may be new to some organizations is the requirement to establish and maintain procedures related to the identifiable significant aspects of goods and services used by the organization, and to communicate these relevant procedures and requirements to suppliers and contractors. Essentially, the organization must ensure that contractor activities performed on site do not conflict with the organization's environmental policy, objectives and targets, and overall EMS. In order to do this, the organization should consider methods of communicating relevant elements of the EMS to contractors. Examples of how this communication can take place are presented in Fig. 13-1.

Supplier Management

In addition to managing its own operations, an organization must ensure that its suppliers and contractors understand the requirements of its EMS, so that they cannot negatively affect the system. This is not to say that the organization can (or should) impose its EMS on the sup-

---

Methods of Communicating to Contractors about the Organization's EMS

✓ Provide information about the EMS in the purchase contract
✓ Provide contractors with handbooks which detail the environmental policy and procedures they are expected adhere to while on site
✓ Require the contractor to review an EMS awareness video before beginning the job
✓ Communicate EMS information in an awareness training session prior to working on site
✓ Perform informal reviews at the contractor's job site and communicate any nonconformances to the contractor procedures

---

Figure 13-1. Examples of methods to communicate relevant elements of the EMS to contractors.

plier or contractor; rather the organization needs to ensure that goods and services provided by the supplier or contractor meet its own requirements. For instance, if the organization has determined that it wants to use asbestos-free parts, then it must make sure that the parts it buys meet this requirement. The supplier may make some parts with asbestos and some parts without. The standard does not intend for the organization to require the supplier to quit making parts with asbestos, only that the organization ensures that the parts it buys from that supplier are asbestos-free. Nonetheless, the organization can choose to buy from suppliers that make only asbestos-free parts.

Operational Control of Quality Seat Belts, Inc. (QSBI)

The environmental operations officer of our hypothetical company, QSBI, performed a gap analysis of this section of the ISO 14001 standard, which is presented in Fig. 13-2. As shown in the figure, operations and activities associated with significant environmental aspects have been identified during the process of identifying significant environmental aspects. A summary of these operations and activities at relevant functions and levels is presented in Table 13-1. The gap analysis confirms that operational procedures are in place, appropriately; however, the environmental operations officer wants a list of these to be maintained at each relevant function and level. In addition, the environmental operations officer has decided to add a question in the environmental questionnaire to request this list from all plant sites.

The gap analysis also revealed that procedures need to be in place for managing significant aspects of goods and services used by QSBI and for communicating relevant procedures about the EMS to contractors, vendors, and suppliers. This procedure, presented in Fig. 13-3, was developed at the corporate level and pertains to services provided by contractors, vendors, and suppliers worldwide.

QSBI's Gap Analysis of Operational Control

Requirement	*QSBI's Current Status*	*Comments*
Operations and activities that are associated with the significant aspects have been identified	Conforms	This list was generated during the process of identifying significant environmental aspects.
Procedures are established and maintained to cover situations where their absence could lead to deviations from the environmental policy and the objectives and targets	Conforms	The Environmental Operations Officer and Environmental Coordinators should maintain a list of these procedures at their respective levels.
The procedures stipulate operating criteria	Conforms	The Environmental Operations Officer and Environmental Coordinators, as appropriate, should review procedures to ensure operating criteria are specified.
Procedures are established and maintained for identifiable significant aspects of goods and services used by QSBI	Does not conform	The Environmental Operations Offficer and Environmental Coordinators, as appropriate, should establish these procedures.
Relevant EMS procedures and requirements are communicated to suppliers and contractors	Partially conforms	A formalized process is needed for communicating to suppliers and contractors about relevant elements of the EMS.

Prepared by John Smith, Environmental Operations Officer   February 25, 1997

Figure 13-2. QSBI's gap analysis for Section 4.4.6 of ISO 14001.

Table 13-1. Summary of QSBI's Operations and Activities
That Have Identified Significant Aspects Associated with Them

Corporate Headquarters	
Activities	Significant aspects
General office activities	Energy use (office building) Aluminum cans and paper recycling

McAllen, Texas Plant	
Activities	Significant aspects
General office activities	Energy use (building use) Aluminum can and paper recycling
Electroplating operations	Chemical use (plating chemicals) Energy use Wastewater discharges Air emissions
Metal shop operations	Chemical use (oils and degreasing agents) Energy use
Test operations	Energy use
Facility operations	Chemical use (water and wastewater treatment chemicals, lawn maintenance chemicals) Energy use (boilers; chillers; building heating, cooling, and lighting) Wastewater discharge (cooling-tower blowdown)
Corrective action	Groundwater remediation

Mexico City Plant	
Activities	Significant aspects
General office activities	Aluminum can and paper recycling
Plastic shop operations	Material consumption (plastic beads) Energy use Nonhazardous waste (scrap plastic)
Assembly operations	Material consumption (plastic, metal, and nylon parts) Nonhazardous waste (plastic pieces, rags, and gloves)
Shipping/receiving/packaging	Material consumption (boxes, pallets, shrink wrap) Nonhazardous waste

Table 13-1. Summary of QSBI's Operations and Activities
That Have Identified Significant Aspects Associated with Them
(*Continued*)

Facility operations	Chemical use (water-treatment chemicals)
	Energy use (boilers; heating coils; building heating, cooling, and lighting)
	Nonhazardous waste (empty, rinsed containers)

Assembly Plants (Milan, Italy and São Paulo, Brazil)	
Activities	Significant aspects
General office activities	Aluminum can and paper recycling
Facility operations	Energy use (building heating, cooling, and lighting)

Procedure Name: Procedure for Operational Control of Contractors
Document Control Number: QSBI Corporate Procedure 4.4.6
Document Owner: John Smith, Environmental Operations Officer

Introduction.

QSBI purchases services from contractors, vendors and suppliers at both the Corporate and plant site levels that could have an effect on its environmental management system (EMS). QSBI is committed to ensuring that these service providers understand the importance of supporting the EMS while on QSBI premises.

Requirements and Responsibilities.

Corporate Headquarters

The Environmental Operations Officer, in conjunction with the Purchasing/Distribution Officer, is responsible for identifying contractors, vendors, and suppliers needed to provide services worldwide related to QSBI's EMS. This includes, but is not limited to, the identification of waste disposal and recycling vendors, suppliers of raw materials, and contractors performing janitorial and maintenance services. As part of the contract, the Purchasing/Distribution Officer is responsible to ensure each contractor, vendor, and/or supplier receives the booklet entitled *Contractor Responsibilities While on QSBI Premises.* This document defines requirements to ensure that the contractor works within QSBI's EMS and is aware of how his/her operations and activities can support the Corporation's environmental policy and objectives and targets.

In addition, the Environmental Operations Officer is responsible for evaluating significant environmental aspects and impacts of services requested and for ensuring that these are in line with QSBI's environmental policy and the objectives and targets. Further, the Environmental Coordinator is responsible for establishing and maintaining a Contractor Audit Checklist. This checklist is to be used to audit contractors to ensure that they meet requirements of the document mentioned above and to ensure that their operations and activities support the EMS.

Plant Site

The Environmental Coordinator, in conjunction with the person charged with responsibility for purchasing at the plant site level (Purchasing Coordinator in McAllen and Mexico City plant sites and the Shipping/Receiving Coordinator in Milan and São Paulo) is responsible for identifying contractors, vendors, and suppliers, as needed, to carry out

Figure 13-3. QSBI's procedure for operational control of contractors.

Procedure Name: Procedure for Operational Control of Contractors
Document Control Number: QSBI Corporate Procedure 4.4.6

activities related to the plant level EMS that are not part of Corporate Headquarters' contracts. As part of the contract, the purchasing representative is responsible to ensure each contractor, vendor, and/or supplier receives the document entitled *Contractor Responsibilities While on QSBI Premises*. This docment defines requirements to ensure that the contractor works within QSBI's EMS and is aware of how his/her operations and activities can support the Corporation's environmental policy and objectives and targets.

In addition, the Environmental Coordinator is responsible for evaluating significant environmental aspects and impacts of services requested, and for ensuring that these are in line with QSBI's environmental policy and the objectives and targets. Further, the Environmental Coordinator is responsible for performing audits of contractors on premises at least quarterly using the Contractor Audit Checklist. Any nonconformances to the above document or support of the EMS must be documented and corrective and/or preventive action taken.

References.

Contractor Responsibilities While on QSBI Premises.
Contractor Audit Checklist.

Records.

Records of contracts.
Records of contractor audits.

Change History.

Date Issued: May 9, 1997.
Last Revision: Not Applicable.

Figure 13-3. (*Continued*) QSBI's procedure for operational control of contractors.

14

Section 4.4.7: Emergency Preparedness and Response

General Requirements

This section of the ISO 14001 standard requires an organization to establish and maintain procedures to (1) identify potential for and respond to accidents and emergency situations, and (2) prevent and mitigate the environmental impacts that may be associated with them. In addition, the organization must review and revise the emergency preparedness and response procedures, where necessary, in particular after the occurrence of accidents or emergency situations. Finally, the standard requires the organization to periodically test these procedures where practicable.

Most organizations that have a mature environmental management system (EMS) have already documented these types of procedures—often in the form of a lengthy and detailed emergency response plan. Examples of what is typically found in an emergency response plan are presented in Table 14-1.

Table 14-1. Examples of Information Typically Found in an
Emergency Response Plan

General Information

- Overview of operations and potential emergency scenarios
- List of applicable regulations
- Maintenance and distribution of plan
- Designated personnel (i.e., names and phone numbers of persons authorized to act as emergency coordinators)

Emergency Notification Information/Procedures

- Information or procedures to effectively notify persons listed in the plan, including facility coordinators and community emergency personnel
- Emergency hotline numbers
- Lists of names and numbers of regulatory agency or other governmental contacts

Reportable Information

- Name and telephone number of person reporting incident
- Name and address of facility
- Time and place of incident
- Name and quantity of material involved
- Extent of injuries
- Possible hazards to human health and/or the environment
- Potential impact outside the facility
- Occurrence of or need for evacuation of local areas

Available Resources Listings

- Spill equipment list, with locations identified
- List of nearby medical facilities, with addresses and phone numbers
- List of nearby fire stations, with addresses and phone numbers
- Poison control center phone number

Emergency-Specific Information/Procedures

- Weather emergencies
- Utility emergencies
- Fire and explosion emergencies
- Chemical release emergencies
- Evacuation procedures

Table 14-1. Examples of Information Typically Found in an Emergency Response Plan (*Continued*)

Site Map
■ Identification of evacuation routes
■ Identification of safety showers and eye-wash stations
■ Identification of fire extinguishers
■ Identification of chemical-spill supply carts

Training and Awareness
■ Legal/regulatory training requirements
■ List of personnel who require training
■ List of qualified training instructors

Quality Seat Belts, Inc. (QSBI) Management of Emergency Preparedness and Response

The environmental operations officer performed a gap analysis of this section of the standard, which is presented in Fig. 14-1. Because QSBI has had emergency procedures in place for many years at the plant site level, the only thing that needed to be done was to place the procedures into the document control system. An example of the McAllen plant's procedure to meet this requirement is presented in Fig. 14-2.

QSBI's Gap Analysis of Emergency Preparedness and Response

Requirement	QSBI's Current Status	Comments
Procedures have been established and maintained to identify potential for and respond to accidents and emergency situations, and for preventing and mitigating the environmental impacts that may be associated with them	Conforms	Although procedures are in place at all plant sites, they need to be put into a document control system.
The procedures are revised, when necessary	Conforms	Procedures are reviewed and revised as necessary.
The procedures are periodically tested, as applicable	Conforms	Procedures are tested at least biennially.

Prepared by John Smith, Environmental Operations Officer   March 11, 1997

Figure 14-1. QSBI's gap analysis for Section 4.4.7 of ISO 14001.

Procedure Name: Emergency Preparedness and Response
Document Control Number: McAllen Plant Procedure 4.4.7
Document Owner: Javier Ramirez, QSBI Environmental Coordinator

Introduction.

The QSBI McAllen Plant has identified the potential for accidents and emergency situations. As a result, emergency preparedness and response is a formalized process that addresses prevention, response, and mitigation of these potential emergencies. This process is detailed in the site's *Emergency Action Plan.*

Requirements and Responsibilities.

The requirements and responsibilities pertaining to emergency preparedness and response are as follows:

a. The Plant Manager has overall responsibility for emergency preparedness and response at the McAllen Plant Site. As such, this person is responsible for providing resources -- both human and financial -- needed to handle an unplanned release or other emergency that could impact the environment. In addition, the Plant Manager is responsible for reviewing and approving the *Emergency Action Plan*, designating an Emergency Coordinator, and ensuring that overall prevention, response, and mitigation measures are adequate and effective.

b. The Environmental Coordinator has been designated as the Emergency Coordinator. As such, this person is responsible for establishing and maintaining the *Emergency Action Plan*, for coordinating activities during emergency situations, and for testing the *Emergency Action Plan* at least biennially. In addition, the Environmental Coordinator is responsible for assessing, initiating, and documenting corrective and preventive actions following emergency situations.

c. The Human Resources/Public Relations Coordinator is responsible to maintain a list of employees assigned to job functions for emergency response. In addition, that person is responsible for coordinating and documenting training to ensure proper response to emergency situations. as well as comply with applicable training requirements. Further, this person is responsible for external communication about accidents and/or incidents, as appropriate.

Figure 14-2. McAllen's procedure for emergency preparedness and response.

Procedure Name: Emergency Preparedness and Response *Cont.*
Document Control Number: McAllen Plant Procedure 4.4.7

References.

Emergency Action Plan.
OSHA 29 CFR 1910.120.
Hidalgo County Local Emergency Plan.

Records.

Documented corrective and preventive actions to emergencies.
Records of biennial tests of the *Emergency Action Plan.*
List of personnel assigned emergency response job functions.
Training records.
Records of external communications about accidents and/or incidents.

Change History.

Date Issued: January 10, 1986.
Last Revision: May 6, 1997 (added Document Control Number and reformatted procedure to
conform with document control format.)

Figure 14-2. (*Continued*) McAllen's procedure for emergency preparedness and
response.

PART 4

Checking and Corrective Action

15

Section 4.5.1: Monitoring and Measurement

Monitoring and Measurement of Key Characteristics

This element of the ISO 14001 standard requires that the organization establish and implement procedures to monitor and measure, on a regular basis, key characteristics of its operations and activities that could have a significant impact on the environment. When conforming to this requirement, the organization must record information to track performance, relevant operational control, and conformance with the organization's environmental objectives and targets.

Although the organization must ensure that its environmental objectives and targets are tracked, this requirement goes beyond that by requiring monitoring and measuring activities for all key characteristics. Examples of key characteristics that might be included in a monitoring and measurement program are presented in Fig. 15-1. Methods to track performance and relevant operational control for each of the key characteristics shown in Fig. 15-1 are presented in Table 15-1.

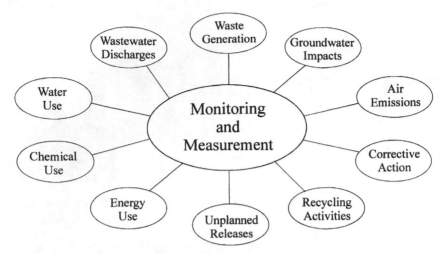

Figure 15-1. Examples of key characteristics to include in a monitoring and measurement program.   (*Source: Adapted from Cascio, Woodside, and Mitchell, 1996.*)

Calibration of Monitoring Equipment

In addition to monitoring and measurement of key characteristics, this element of the standard requires monitoring equipment to be calibrated and maintained, with calibrations recorded, per the organization's procedures. The standard leaves it up to the organization to determine which pieces of monitoring equipment and calibration frequencies will be specified in the procedures. Certainly, all monitoring equipment used for compliance to regulations and operational control should be included.

Periodic Evaluation of Compliance with Legislation and Regulations

The final requirement defined in this section of the standard specifies that the organization must put in place a documented procedure for periodically evaluating compliance with relevant environmental legislation and regulations. Unlike the EMS audit, which is discussed in Chap. 18, this requirement does not necessarily require an audit, per se, and the evaluation does not have to be performed by an independent

Table 15-1. Methods to Track Performance and Relevant Operational Control for Key Characteristics

Key characteristic	Methods to track performance and operational control
Air emissions	Periodic stack sampling; on-line monitoring; emissions trending; measurement of exhaust fan rates; frequent calibration of monitoring equipment; balancing of duct flows; mass-balance calculations
Treatment efficiencies	Influent versus effluent comparisons of specific parameters; emissions trending; comparison to or use of documented efficiency factors; waste generation from process(es); frequent calibration of monitoring equipment; mass-balance calculations
Recycling activities	Year-to-year comparison of amounts or percentage of waste streams recycled, using activity index; year-to-year comparison of revenue generated and cost avoided
Unplanned releases	Year-to-year trend of number of unplanned releases; year-to-year trend of severity of unplanned releases; cost or potential costs of cleanup versus cost of capital equipment
Energy use	Annual tracking of energy usage; cost of installing energy-efficient equipment versus savings over time; periodic audits of areas to ensure specified energy conservation measures are being deployed
Chemical use	Year-to-year chemical use trends, using an activity index; review of purchasing records to determine what areas are using the most chemicals; reduction of targeted chemicals such as toxics or hazardous chemicals; audits of processes to ensure that specified chemical conservation measures are being deployed; mass-balance calculations
Water use	Year-to-year water use trends, using an activity index; cost of water conservation equipment versus savings over time; audits of processes to ensure that specified water conservation measures are being deployed
Wastewater discharges	Effluent sampling to ensure permit parameters are being met; emissions trending; frequent calibration of monitoring equipment; mass-balance calculations
Waste generation	Year-to-year generation trends, using an activity index; disposal and treatment costs; number of incorrectly shipped waste drums or packages, including errors in shipping papers; mass-balance calculations
Stormwater discharges	Effluent sampling to ensure permit parameters are being met; emissions trending; visual inspection of runoff; frequent calibration of monitoring equipment; mass-balance calculations

person. Examples of what might be reviewed when conforming to this requirement are presented in Fig. 15-2.

Meeting This Element of the Standard at Quality Seat Belts, Inc. (QSBI)

When evaluating how to conform to the requirements of this element of the standard, the environmental operations officer at QSBI—our hypothetical company—first performed a gap analysis, which is presented in Fig. 15-3. As shown in the figure, monitoring and measurement procedures needed to be documented at the corporate and plant site levels. Further, the method for evaluating environmental compliance at the plant sites needed to be proceduralized at that level. These procedures were developed by the environmental operations officer and the environmental coordinator, as appropriate, and are presented for corporate and the Mexico City plant in Figs. 15-4 to 15-6, respectively. Calibration procedures for monitoring equipment already exist at the plant sites.

Elements That Might Be Reviewed When Performing a Compliance Evaluation

✓ Environmental permit limits and corresponding data demonstrating compliance with these limits
✓ Procedures identifying training requirements that are based on legislation/regulation and evidence that the training has taken place
✓ Reporting requirements and evidence that the proper reports have been submitted
✓ Inspection requirements and evidence that the inspections have taken place
✓ Labeling requirements and evidence that proper labeling is being executed
✓ Records retention requirements and evidence that the records are maintained according to legal specification
✓ Waste-handling requirements and evidence that these requirements are being met
✓ Corrective action requirements and evidence that the corrective action is being managed properly

---

Figure 15-2. Examples of what might be reviewed during a compliance evaluation.

QSBI's Gap Analysis of Monitoring and Measurement

Requirement	QSBI's Current Status	Comments
Procedures are established to monitor and measure, on a regular basis, key characteristics	Partially conforms	Monitoring and measurement occurs, but formalized procedures should be established at the Corporate and plant site levels.
There are monitoring equipment calibration procedures	Conforms	These procedures exist at the plant site level.
There are documented procedures for periodically evaluating compliance	Partially conforms	A semi-annual evaluation occurs at McAllen and Mexico City Plants; annual evaluation occurs at Milan and São Paulo Plants. No sites have documented procedures for this activity.

Prepared by John Smith, Environmental Operations Officer   March 18, 1997

Figure 15-3. QSBI's gap analysis for Section 4.5.1 of ISO 14001.

However, they needed to be reviewed and updated to include the document change history and, as appropriate, new compliance requirements under the Clean Air Act Amendments. An example of the calibration procedure for the McAllen, Texas plant is presented in Fig. 15-7.

References

Cascio, Joseph, Gayle Woodside, and Philip Mitchell, *ISO 14000 Guide,* McGraw-Hill, New York, 1996.

Procedure Name: Monitoring and Measurement
Document Control Number: QSBI Corporate Procedure 4.5.1
Document Owner: John Smith, Environmental Operations Officer

Introduction.

Monitoring and measurement of key characteristics, in particular for key characteristics related to significant environmental aspects and objectives and targets, is performed at the Corporate Headquarters' level through the Environmental Questionnaire.

Requirements and Responsibilities.

The Environmental Operations Officer sends out the Environmental Questionnaire to all plant sites quarterly. This questionnaire requests data about the following: aluminum can and paper recycling efforts at all plant sites; chemical use and waste water discharges (McAllen plant site only); and material consumption and nonhazardous waste (Mexico City plant site only). Based on the data, the Environmental Operations Officer documents environmental performance trends for the corporation.

References.

Environmental Questionnaire.

Records.

Completed Environmental Questionnaires from the plant sites.

Document History.

Date Issued: May 8, 1997.
Last Revision: Not Applicable.

Figure 15-4. Corporate headquarters' procedure for monitoring and measurement.

Procedure Name: Monitoring and Measurement
Document Control Number: Mexico City Plant Procedure 4.5.1A
Document Owner: Guadalupe Martinez, QSBI Environmental Coordinator

Introduction.

The Mexico City Plant Site monitors and measures key characteristics on a regular basis. These measurements help the plant site maintain operational control and provide data on progress toward achieving its objectives and targets.

Requirements and Responsibilities.

The Mexico City Plant monitors and/or measures its key characteristics as follows:

Key Characteristic	*Method of Monitoring and/ or Measurement*	*Person Responsible*
Energy use	Continuous metering of energy at utility plant; monthly trending	Environmental Coordinator
Aluminum can and paper recycling	Track amounts sent to recycling vendors monthly (shipping papers)	Supervisor for Shipping and Receiving
Material consumption (plastics)	Track kilograms of plastic monthly (receipts)	Supervisor for Shipping and Receiving
Nonhazardous waste (plastics)	Track kilograms of waste shipped monthly (shipping papers)	Supervisor for Shipping and Receiving
Heat generated in plastics shop	Track temperature twice daily (0600 and 1400 hours) using mounted thermometer	Supervisor for Plastics shop

Figure 15-5. Mexico City's procedure for monitoring and measurement.

Procedure Name: Monitoring and Measurement *Cont.*
Document Control Number: Mexico City Plant Procedure 4.5.1A

The Environmental Coordinator compiles the above data into a monthly data tracking report. Quarterly, the Environmental Coordinator submits a summary of data to the Environmental Operations Officer through the Environmental Questionnaire.

Reference.

QSBI Corporate Procedure 4.5.1.

Records.

Monthly trends of energy use at facilities plant.
Shipping papers for recycling shipments of aluminum cans and paper.
Records of plastic materials received.
Shipping papers for nonhazardous plastic waste.
Monthly data tracking report.
Environmental Questionnaire.

Document History.

Document Issued: May 22, 1997.
Last Revision: Not Applicable.

Figure 15-5. (*Continued*) Mexico City's procedure for monitoring and measurement.

Procedure Name: Periodically Evaluating Compliance with Environmental Legislation and Regulations
Document Control Number: Mexico City Plant Procedure 4.5.1B
Document Owner: Guadalupe Martinez, QSBI Environmental Coordinator

Introduction.

The QSBI Mexico City Plant Site is committed to complying with applicable environmental legislation and regulations. Thus, there is a process for periodically evaluating compliance of the manufacturing and assembly activities and operations.

Requirements and Responsibilities.

The Environmental Coordinator evaluates the activities and operations of the Mexico City Manufacturing and Assembly Plant with respect to legal compliance on a semiannual basis, using a compliance checklist that covers all relevant legislation and regulations. Any compliance nonconformance found during the evaluation is noted on the checklist, and corrective and preventive action is taken and documented. Any serious nonconformance identified by the Environmental Coordinator that could result in fines or in negative publicity is communicated to the Corporate Environmental Operations Officer and the Plant Manager immediately. A summary of the compliance evaluation is prepared by the Environmental Coordinator and submitted to the Corporate Environmental Operations Officer.

References.

None.

Records.

Records of completed evaluations, nonconformances, and corrective and preventive action. Compliance evaluation summary.

Document History.

Date Issued: March 3, 1997.
Last Revision: Not Applicable.

Figure 15-6. Mexico City's procedure for periodically evaluating compliance with environmental legislation and regulations.

Procedure Name: Equipment Calibration
Document Control Number: McAllen Plant Procedure 4.5.1C
Document Owner: Javier Ramirez, QSBI Environmental Coordinator

Introduction.

Monitoring and measurement equipment essential for performance tracking and operational control of the EMS is calibrated according to this procedure.

Requirements and Responsibilities.

The following defines the calibration schedule for monitoring and measurement equipment that is part of the environmental management system at QSBI's McAllen Plant.

Electroplating Area

Equipment:	Calibrated:	Responsible Person:
pH sensor for chrome bath	daily	Plating Supervisor
Stalagmometer for chrome bath	daily	Plating Supervisor
Conductivity meter for caustic bath	weekly	Plating Supervisor
Anemometer for plating bath exhaust hoods	monthly	Maintenance Supervisor

Wastewater Treatment Process

Equipment:	Calibrated:	Responsible Person:
High-level indicator on equalization tank	weekly	Maintenance Supervisor
pH sensor on neutralization tank #1	daily	Plating Supervisor
pH sensor on neutralization tank #2	daily	Plating Supervisor
Flow rate sensor on discharge tank	monthly	Maintenance Supervisor

Figure 15-7. McAllen's procedure for monitoring equipment calibration.

Procedure Name: Calibration Procedure *Cont.*
Document Control Number: McAllen Plant Procedure 4.5.1C

Groundwater Monitoring Program (Corrective Action)

Equipment:	Calibrated:	Responsible Person:
Temperature probe	monthly	Environmental Coordinator
Conductivity meter	monthly	Environmental Coordinator
Groundwater level indicator	yearly	Environmental Coordinator
Groundwater samples*	monthly	Environmental Coordinator

*Quality control calibrations are conducted by vendor laboratory for each monthly sample.

References.

None.

Records.

Calibration records for monitoring and measurement equipment identified in this procedure.

Document History.

Date Issued: January 12, 1993.
Last Revision: June 2, 1997. (Added stalagmometer readings to comply with Maximum Achievable Control Technology (MACT) standard.)

Figure 15-7. (*Continued*) McAllen's procedure for monitoring equipment calibration.

16

Section 4.5.2: Nonconformance and Corrective and Preventive Action

General Requirements

ISO 14001 requires that the organization establish and maintain procedures for defining responsibility and authority for handling and investigating nonconformances, for taking action to mitigate any environmental impacts, and for initiating and completing corrective and preventive action. The corrective or preventive action taken to eliminate the causes of nonconformances should be appropriate to the magnitude of the problems encountered. Finally, the standard requires the organization to make changes to procedures, as necessary, as a result of corrective and preventive action.

This requirement may sound as though it's directed at regulatory noncompliance and/or emergency situations, and certainly these types of nonconformances may warrant comprehensive and far-reaching corrective and preventive actions. However, this section of the standard can also apply to any nonconformances identified within the environmental management system (EMS). Examples of types of nonconformances that might be identified within an organization's EMS are presented in Fig. 16-1.

Types of Nonconformances That Might Be Identified Within the EMS

✓ Activities or operations do not support the environmental policy

✓ Employees whose job functions could affect legal requirements do not have access to these requirements

✓ Significant environmental impacts have not been defined

✓ Views of interested parties were not considered when setting objectives and targets

✓ Environmental management program does not specify responsible person

✓ Defined roles, responsibilities, and authorities have not been communicated to relevant employees

✓ Training plan is not being followed

✓ Relevant external communications are not documented

✓ Direction to those documents that interact with the core elements of the EMS is not provided

✓ Obsolete documents were not removed promptly from points of use

✓ Contractors were not aware of organization's procedures that related to the EMS

✓ Emergency plan was not tested as prescribed in procedures

✓ Equipment calibration schedule was not followed

✓ Responsibility for handling nonconformances to the EMS was not defined

✓ Records were not easily retrievable

✓ The EMS audit schedule was not defined

✓ The management review was not documented

Figure 16-1. Examples of types of nonconformances that might be identified within an EMS.

Management of Nonconformances at Quality Seat Belts, Inc. (QSBI)

At QSBI, the environmental operations officer performed a gap analysis of this section of the standard, and the results are presented in Fig. 16-2. As shown in the figure, QSBI has procedures in place for corrective and preventive action for regulatory noncompliance emergency situations, but not for handling and investigating nonconformances within other areas of the EMS. Thus, the environmental operations officer established and developed such a procedure, which is presented in Fig. 16-3.

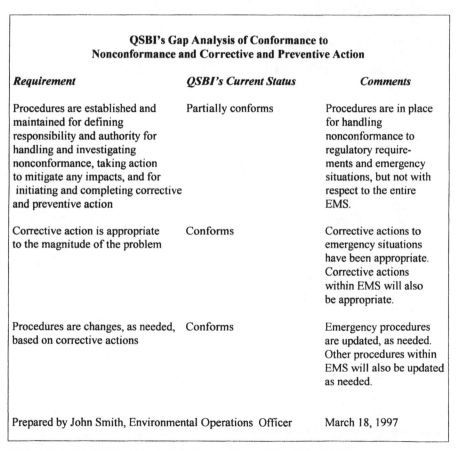

QSBI's Gap Analysis of Conformance to
Nonconformance and Corrective and Preventive Action

Requirement	QSBI's Current Status	Comments
Procedures are established and maintained for defining responsibility and authority for handling and investigating nonconformance, taking action to mitigate any impacts, and for initiating and completing corrective and preventive action	Partially conforms	Procedures are in place for handling nonconformance to regulatory requirements and emergency situations, but not with respect to the entire EMS.
Corrective action is appropriate to the magnitude of the problem	Conforms	Corrective actions to emergency situations have been appropriate. Corrective actions within EMS will also be appropriate.
Procedures are changes, as needed, based on corrective actions	Conforms	Emergency procedures are updated, as needed. Other procedures within EMS will also be updated as needed.
Prepared by John Smith, Environmental Operations Officer		March 18, 1997

Figure 16-2. QSBI's gap analysis for Section 4.5.2 of ISO 14001.

Procedure Name: Nonconformance and Corrective and Preventive Action
Document Control Number: QSBI Corporate Procedure 4.5.2
Document Owner: John Smith, Environmental Operations Officer

Introduction.

QSBI is committed to taking corrective and preventive action to mitigate nonconformances
identified within the environmental management system (EMS). QSBI has taken such
measures over the years when emergency situations have occurred, and is expanding this
practice to include nonconformance to all elements of the EMS. Corrective and preventive
action will be appropriate to the magnitude of the problem (potential or actual) identified
and will be in line with the environmental policy and environmental impact of the
nonconformance.

Requirements and Responsibilities.

Requirements and responsibilities for nonconformance and corrective and preventive action
at the Corporate and plant site levels are as follows:

Corporate Headquarters

The Environmental Operations Officer is responsible for handling any corporate-wide
nonconformance to the EMS identified during the EMS audit process, during review of
Environmental Questionnaires, and/or through other means. This person will investigate the
nonconformance using root cause analysis and will develop a plan for corrective and
preventive action. Should human or financial resources be needed to initiate the plan, these
must be authorized by the President of QSBI. Once the corrective and preventive action plan
is initiated, the Environmental Operations Officer will document and track all such actions
to closure.

Figure 16-3. QSBI's procedure for managing nonconformance and corrective and
preventive action.

Procedure Name: Nonconformance and Corrective and Preventive Action *Cont.*
Document Control Number: QSBI Corporate Procedure 4.5.2

Plant Sites

The Environmental Coordinator is responsible for handling any site-specific
nonconformance to the EMS identified during the EMS audit process, during compilation of
the Environmental Questionnaire, through compliance and/or contractor audits, and/or
through other means. This person will investigate the nonconformance using root cause
analysis and will develop a plan for corrective and preventive action. Should human or
financial resources be needed to initiate the plan, these must be authorized by the Plant
Manager. Once the corrective and preventive action plan is initiated, the Environmental
Coordinator will document and track all such actions to closure.

References.

QSBI Corporate Procedure 4.5.4 (EMS Audit).
Environmental Questionnaire.
Compliance audit checklist.
Contractor audit checklist.

Records.

Records of identification of nonconformances.
Records of corrective and preventive actions.

Change History.

Date Issued: June 2, 1997.
Last Revision: Not Applicable.

Figure 16-3. (*Continued*) QSBI's procedure for managing nonconformance and
corrective and preventive action.

17

Section 4.5.3: Records

General Requirements

Records are different from documents. Documents include procedures, instructions, manuals, and other forms of documentation that are used to manage the environmental management system (EMS). Records, on the other hand, are evidence that something has been accomplished (i.e., inspections, equipment calibration, and training). ISO 14001 defines management of both documents and records, with control of documents being defined in Section 4.4.5 (see Chap. 11) and maintenance of records being defined in this section. In addition, the standard requires the organization to identify core elements of the EMS and the documents that interact with them; these requirements are outlined in Section 4.4.4 (see Chap. 10).

Like many other sections of the standard, this section requires the organization to establish and maintain procedures—this time for the identification, maintenance, and disposition of records. The standard delineates that the organization must include training records and the results of the EMS audit and management reviews. Actually, the organization's environmental records typically will encompass much more than these. Examples of types of environmental records for which the above-mentioned procedures might apply are presented in Fig. 17-1. Further, this section requires that environmental records be:

- Legible
- Identifiable and traceable to the activity, product, or service involved

---

Examples of Environmental Records

✓ Applicable governmental regulations

✓ List of significant environmental aspects

✓ Environmental permits

✓ Training records

✓ Process hazard assessments

✓ Emissions modeling records

✓ Information on product attributes

✓ External environmental reports

✓ Inspection, maintenance, and calibration records

✓ Records of contractor activities on premises

✓ Incident and corrective action reports

✓ Records of testing of emergency procedures

✓ Records of compliance and EMS audit results

✓ Records of management review

Figure 17-1. Examples of types of environmental records that an organization might keep.

- Easily retrievable
- Protected against damage, deterioration, or loss
- Retained per established and recorded retention times

This requirement should pose no real burden on the mature EMS, although procedures for managing records may need to be expanded.

Records Management at Quality Seat Belts, Inc. (QSBI)

The environmental operations officer performed a gap analysis of QSBI's conformance to this section of the standard, which is presented in Fig. 17-2. As shown in the figure, QSBI has identified its environmental records and manages them adequately but does not have a formalized procedure to do so. Thus the environmental operations officer established a procedure to use throughout the corporation. This procedure is presented in Fig. 17-3.

QSBI's Gap Analysis for Records		
Requirement	*QSBI's Current Status*	*Comments*
Procedures are established and maintained for the identification, maintenance, and disposition of records	Partially conforms	Environmental records have been identified, but formal procedures for managing these are not established.
Environmental records are legible, identifiable and traceable to the activity, product or service	Conforms	Records are identified in the EMS Documentation Hierarchy.
Environmental records are stored and maintained in such a way as to be readily retrievable and protected against damage, deterioration, or loss	Conforms	Records are kept in fireproof cabinets at Corporate Headquarters and at Plant Sites.
Prepared by John Smith, Environmental Operations Officer		March 18, 1997

Figure 17-2. QSBI's gap analysis for Section 4.5.3 of ISO 14001.

Procedure Name: Records
Document Control Number: QSBI Corporate Procedure 4.5.3
Document Owner: John Smith, Environmental Operations Officer

Introduction.

QSBI's environmental records are identified in the EMS Documentation Hierarchy. These records are legible, identifiable and traceable to the activity, product, or service. They are stored and maintained in such a way as to be readily retrievable and protected against damage, deterioration, or loss.

Requirements and Responsibilities.

Responsibilities for managing records are distributed at the Corporate and Plant Site levels, as follows:

Corporate Headquarters

The Environmental Operations Officer is responsible for establishing and maintaining the EMS Documentation Hierarchy, which identifies QSBI's environmental records. Record retention times for the various documents are established as follows:

a. Training records are maintained for 5 years
b. Audit results are maintained for 3 years
c. Management reviews are maintained for 3 years
d. Operational data, including monitoring and measurement data, is kept for 5 years
e. Equipment calibration and maintenance records are maintained for 5 years
f. Inspection records are maintained for 3 years
g. Environmental Questionnaires are maintained for 5 years
h. All other records are maintained for 1 year

At Corporate Headquarters, the Environmental Operations Officer is responsible for ensuring that the records are stored for the prescribed retention time and are maintained appropriately.
The records will be stored in a fireproof cabinet(s) under the control of the Environmental Operations Officer.

Figure 17-3. QSBI's procedure for maintaining records.

Procedure Name: Records
Document Control Number: QSBI Corporate Procedure 4.5.3

Plant Sites

The Environmental Coordinator is responsible for ensuring that plant site records are stored according to the prescribed retention time and are maintained in a proper manner. The records will be stored in a fireproof cabinet(s) under the control of the Environmental Coordinator.

Reference.

EMS Documentation Hierarchy.

Records.

Environmental Records.

Change History.

Date Issued: June 6, 1997.
Last Revision: Not Applicable.

Figure 17-3. (*Continued*) QSBI's procedure for maintaining records.

18

Section 4.5.4: Environmental Management System Audit

General Requirements

Environmental management system (EMS) audits are required by ISO 14001 to assess whether or not the EMS conforms to planned arrangements for the EMS—including conformance to ISO 14001—and has been properly implemented and maintained. Unlike compliance audits, these audits specifically address the EMS as it is established, maintained, and implemented in conformance to the standard.

The standard requires that the organization establish and maintain procedures and a program(s) for EMS auditing. ISO issued a standard pertaining to general principles of environmental auditing in late 1996, which is delineated in ISO 14010. Although this standard is not part of ISO 14001 and is not required for use in any way, it is helpful for those who are novices with respect to audit processes. A summary of the contents of this ISO standard is presented in Table 18-1.

The EMS audit procedures must cover at least the following:

- Audit scope
- Audit frequency

Table 18-1. Summary of the Contents of ISO 14010:
Guidelines for Environmental Auditing—General Principles

Scope
■ Applicable to all types of environmental audits

Definitions
■ Defines key terms pertaining to environmental auditing including audit conclusion, audit criteria, audit evidence, and audit findings

Requirements for an Environmental Audit
■ There is sufficient and appropriate information about the subject matter of the audit ■ There are adequate resources to support the audit process ■ There is adequate cooperation from the client (auditee)

General Principles
■ Objectives and scope—the audit should be based on the objectives defined by the client ■ Objectivity, independence, and competence—the members of the audit team should be independent and should possess an appropriate combination of knowledge, skills, and experience to carry out the audit ■ Due professional care—auditor should use care, diligence, skill, and judgment ■ Systematic procedures—documented and well-defined methodologies should be used to enhance consistency and reliability ■ Audit criteria, evidence, and findings—criteria should be agreed upon between the lead auditor and the client; appropriate information should be collected, analyzed, interpreted, and recorded; audit evidence should be of such quality and quantity that competent environmental auditors working independently of each other will reach similar audit findings ■ Reliability of audit findings and conclusions—the desired level of confidence and reliability in audit findings and conclusions should be provided ■ Audit report—a copy should be given to the client and should include items agreed upon by the client and the lead auditor

- Audit methodologies
- Responsibilities and requirements for conducting audits and reporting results

The frequency of the audits is to be based on the environmental importance of the activity being audited and the results of previous audits. Results of the audits should be provided to management for review. An example of an EMS audit methodology, with EMS requirements, audit questions, and types of objective evidence that would be acceptable to an auditor is presented in Appendix B.

Quality Seat Belts, Inc. (QSBI) EMS Audit Procedures and Program

The environmental operations officer performed a gap analysis of QSBI's conformance to this section of ISO 14001, which is presented in Fig. 18-1. As shown in the figure, QSBI needs to establish procedures and programs for the EMS audit. These were developed by the environmental operations officer and are presented in Figs. 18-2 and 18-3, respectively. The audit methodology selected by QSBI is found in Appendix B.

QSBI's Gap Analysis of Environmental Management System Audit

Requirement	QSBI's Current Status	Comments
Procedures and programs are established and maintained to periodic EMS audits	Does not conform	An environmental compliance program and contractor audit procedure is in place. Procedures and programs need to be developed for EMS audits.
The audit program, including the schedule, is based on the environmental importance of the activity concerned and the results of previous audits	Does not conform	No EMS audits have been conducted to date.
The audit procedures cover the audit scope, the frequency and methodology, and the responsibilities and requirements for conducting audits and reporting results	Does not conform	Procedures need to be written.
Prepared by John Smith, Environmental Operations Officer		March 18, 1997

Figure 18-1. QSBI's gap analysis for Section 4.5.4 of ISO 14001.

Procedure Name: Environmental Management System (EMS) Audit
Document Control Number: QSBI Corporate Procedure 4.5.4A
Document Owner: John Smith, Environmental Operations Officer

Introduction.

QSBI is committed to assuring that its EMS functions properly. In order to do this, the EMS must be audited by an independent team of auditors. This audit occurs at the Corporate Headquarters and at the Plant Sites.

Requirements and Responsibilities.

Requirements and responsibilities for auditing the EMS at Corporate Headquarters and at the Plant Sites are as follows:

Corporate Headquarters

a. The Environmental Operations Officer is responsible for initiating the EMS audit process at Corporate Headquarters. The audit scope, expected frequency, and methodology is determined by this person and is established in the Environmental Management System (EMS) Audit Program -- which is defined in QSBI Corporate Document 4.5.4B. It is the responsibility of the Environmental Operations Officer to be the interface between Corporate Headquarters and the audit team.

b. The EMS audit team consists of Corporate Legal Counsel (lead auditor) and one or more Plant Site Environmental Coordinators. It is the responsibility of the lead auditor to conduct the audit in accordance with the specified methodology and to summarize the findings in an audit report and present it to the President of QSBI.

Plant Sites

a. The Environmental Operations Officer is responsible for initiating the EMS audit process at the Plant Sites. The audit scope, expected frequency, and methodology is determined by this person and is established in the Environmental Management System (EMS) Audit Program -- which is defined in QSBI Corporate Document 4.5.4B.

Figure 18-2. QSBI's procedure for auditing the EMS.

Procedure Name: Environmental Management System (EMS) Audit *Cont.*
Document Control Number: QSBI Corporate Procedure 4.5.4A

b. It is the responsibility of the Plant Site Environmental Coordinator to be the interface between the plant site and the audit team.

c. The Audit team will consist of Corporate Legal Counsel (lead auditor) and one or more of the following: the Environmental Operations Officer and/or an independent Environmental Coordinator. It is the responsibility of the lead auditor to conduct the audit in accordance with the specified methodology and to summarize the findings in an audit report and present it to the Plant Manager.

Reference.

QSBI Corporate Document 4.5.4B -- Environmental Management System Audit Program.

Records.

Records of audit results.

Change History

Date Issued: January 5, 1998.
Last Revision: Not Applicable.

Figure 18-2. (*Continued*) QSBI's procedure for auditing the EMS.

Environmental Management System (EMS) Audit Program
Document Control Number: QSBI Corporate Document 4.5.4B
Document Owner: John Smith, Environmental Operations Officer

EMS Audit Scope.

The EMS audit program covers QSBI's Corporate Headquarters and all QSBI Plant Sites. All elements of the EMS will be audited, with special emphasis placed on the following:

- Support of the environmental policy by Corporate Headquarters and the Plant Sites and communication of such to employees and contractors, where relevant
- Understanding of significant environmental aspects
- Understanding of objectives and targets at all relevant levels
- Training and awareness methods to ensure that employees understand how their job functions impact the environment and the overall EMS, including the environmental policy and objectives and targets
- Communication about the EMS at all relevant levels
- Compliance and contractor audits
- Adherence to operational procedures
- Handling of nonconformance and corrective and preventive action

EMS Audit Frequency.

The EMS at Corporate Headquarters will be audited at least annually, and more frequently as systemic nonconformances warrant.

The EMS at the Plant Sites will be audited as follows:

McAllen, Texas and Mexico City Plant Sites will be audited semiannually since their activities and operations could potentially have the greatest impact on the environment.

Milan, Italy and São Paulo, Brazil will be audited annually since their activities and operations pose a lesser potential for environmental impact than the aforementioned.

Figure 18-3. QSBI's EMS audit program.

Environmental Management System (EMS) Audit Program *Cont.*
Document Control Number: QSBI Corporate Document 4.5.4B

Results of Audits

EMS audit results at Corporate Headquarters will be reviewed by Corporate Legal Counsel
and the President of QSBI to determine if the scope and/or frequency of the audits needs to
be changed.

EMS audit results at the Plant Sites will be reviewed by the Environmental Operations
Officer to determine if the scope and/or frequency of the audits needs to be changed.

Responsibilities and Requirements for Auditors.

The Corporate Legal Counsel will act as the lead auditor on all EMS audits. This person will
follow typical audit protocol. Other audit team members can consist of the Environmental
Operations Officer and/or Environmental Coordinator(s), so long as they are independent of
the site being audited.

Reference.

QSBI Corporate Procedure 4.5.4A.

Document History.

Date Issued: January 6, 1998.
Last Revision: Not Applicable.

Figure 18-3. (*Continued*) QSBI's EMS audit program.

PART 5

Management Review and Putting It All Together

INPUT TO MGMT REVIEW
ISO 14001: 2004

- RESULTS OF INTERNAL AUDITS AND EVALUATIONS OF COMPLIANCE W/ LEGAL AND OTHER REQUIREMENTS

- COMMUNICATIONS (AND COMPLAINTS) FROM EXTERNAL PARTIES

- ENVIRONMENTAL PERFORMANCE OF ORGANIZATION

- STATUS OF OBJECTIVES & TARGETS

- STATUS OF CORRECTIVE ACTIONS

- FOLLOW UP FROM PREVIOUS MGMT REVIEWS

- RECOMMENDATIONS FOR IMPROVEMENT

19
Section 4.6: Management Review

General Requirements

The final element of the ISO 14001 standard is that of management review. This section requires top management of the organization to review the environmental management system (EMS) at specified intervals to ensure its continuing suitability, adequacy, and effectiveness. Although each organization can determine its own meaning for this collection of terms, some examples of attributes of an EMS that could be considered suitable, adequate, and effective include:

- The EMS addresses all activities, products, and services of the organization, including any recent changes in these.
- The EMS does not contain any systemic flaws.
- The EMS provides the framework for continual improvement and prevention of pollution.
- Legal/regulatory requirements are consistently being met.
- Objectives and targets are in line with the organization's environmental policy and have been consistently met.
- The results of EMS audits show that the organization is in conformance with its planned arrangements.

The standard further specifies that the management review process should ensure that necessary information is collected to allow management to carry out the evaluation. Examples of types of information that

---

Examples of Information That Might Be Included During Management Review

✓ Environmental progress toward achieving objectives and targets

✓ Data about key characteristics

✓ Plans for technological upgrades

✓ Any accidents or incidents that had an adverse impact on the environment and corrective and preventive action taken

✓ Permit compliance records

✓ Anticipated changes in legal requirements

✓ Concerns of interested parties

✓ Results of EMS audits

✓ Changes in activities, products, or services that might require changes in the EMS

Figure 19-1. Examples of information that might be included during the management review process.

might be included in the management review process are presented in Fig. 19-1.

In addition to reviewing information about the EMS, the standard requires that top management address the possible need for changes to policy, objectives and targets, and other elements of the EMS in light of the EMS audit results, any changing circumstances, and the commitment to continual improvement. It is expected that after the management review, the EMS will be revised to reflect the outcome of the review process. In particular, it is expected that the EMS itself will continually improve, thereby enhancing improvement of environmental performance. Examples of what might be considered continual improvements of the EMS are presented in Fig. 19-2.

Management Review at Quality Seat Belts, Inc. (QSBI)

The environmental operations officer at QSBI performed a gap analysis of QSBI's current status with respect to this element of the standard, which is presented in Fig. 19-3. As shown in the figure, QSBI has semi-

---

**Examples of Changes to the EMS That Might
Demonstrate Commitment to Continual Improvement**

✓ Changes to the environmental policy to reflect additional commitments

✓ Identification of additional significant environmental aspects

✓ Changes to objectives and targets which will lead to improved environmental performance

✓ Technology upgrades, such as double containment, improvements to the wastewater treatment plant, and chemical-resistant floor coatings

✓ Enhanced training program

✓ Enhanced external communications

✓ Monitoring and measurement of additional parameters

✓ More frequent tests of emergency preparedness and response procedures

✓ Enhanced document control

✓ Enhanced operational controls, such as installation of alarms, automation of chemical feed systems, use of containers that are easier and/or safer to transport

✓ Investigation of alternative fuel sources

✓ Evaluation of a work-at-home program during ozone action days

✓ Enhanced preventive maintenance program

Figure 19-2. Examples of revisions to the EMS that might demonstrate commitment to continual improvement.

annual reviews with top management on environmental compliance and progress toward goals listed in the total quality environmental management plan, which has now been superseded by the environmental management program (see Chap. 7). Since these reviews do not address the entire EMS, they will need to be modified to include EMS information, and they should be documented.

The environmental operations officer decided to implement the management review process after an EMS audit was performed at corporate headquarters and the plant sites. Since the president of QSBI was familiar with compliance audits, the environmental operations officer used the model presented in Chap. 1 to pictorially show the interrelation of

QSBI's Gap Analysis of Management Review

Requirement	QSBI's Current Status	Comments
Top management reviews the EMS at determined intervals to ensure its continuing suitability, adequacy, and effectiveness	Partially conforms	Top management reviews QSBI's compliance and progress toward meeting requirements in the Total Quality Environmental Management Plan (TQEMP).
The management review process ensures necessary data is collected for top management to review the EMS	Partially conforms	Top management needs to be provided with more data about the EMS.
Management review addresses the need for changes to the policy, objectives and targets and other elements of the EMS	Partially conforms	Currently addresses goals in TQEMP. Need to add environmental policy and other elements of the EMS.
The management review process is documented	Does not conform	Need to document management reviews with note(s) to file.
Prepared by John Smith, Environmental Operations Officer		March 18, 1997

Figure 19-3. QSBI's gap analysis for Section 4.6 of ISO 14001.

legal/regulatory compliance with the EMS as a whole. This model is presented in Fig. 19-4.

Once all pertinent information needed for the management review was collected, the environmental operations officer asked the plant site environmental coordinators to meet with their respective plant managers to review the status of the EMS, including plant-specific information. Once these reviews were complete, the environmental operations officer met with the president of QSBI to review the corporatewide EMS in terms of its suitability, adequacy, and effectiveness. The documented memo to file about this review is presented in Fig. 19-5.

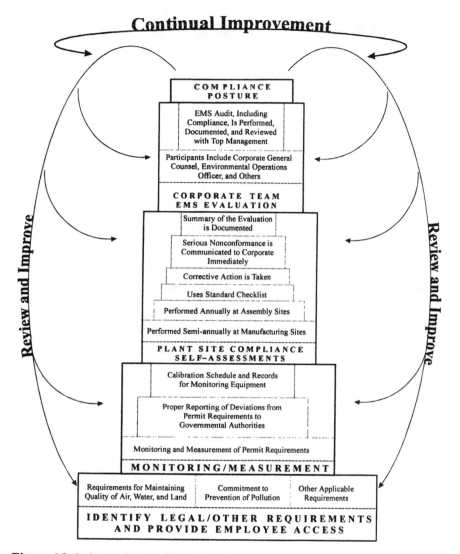

Figure 19-4. Interrelation of legal/regulatory compliance with the EMS.

Memo to File
Date: October 30, 1997

John Smith, Environmental Operations Officer, Quality Seat Belts, Inc., (QSBI)
Re: Environmental Management System (EMS) Review with James Wilson, President, QSBI

James Wilson reviewed QSBI's EMS to ensure that it is suitable, adequate, and effective.
The following information was reviewed:

1. The list of significant environmental aspects for QSBI.
2. The environmental management program (formerly the Total Quality Environmental
Management Plan), with specifics about the process for meeting objectives and targets.
3. Highlights of Plant Sites' environmental achievements, per the Environmental
Questionnaire.
4. A list of new procedures developed to align QSBI's EMS with ISO 14001.
5. Results of EMS audits of Plant Sites and Corporate, performed during September and
early October, 1997.

The following suggestions were made by James Wilson to ensure continual improvement of
the EMS:

1. Review time frames for objectives and targets to assess whether any activities can be
completed earlier.
2. Consider adding water usage to the McAllen significant aspects.
3. Enhance communication of the revised environmental policy both internally and
externally.
4. Consider a process for external communication of the corporation's significant aspects and
environmental management program.
5. Review EMS audit methodology for potential improvements, and consider enhancements
to the overall EMS audit process.

As a result of the review of QSBI's EMS and since QSBI's activities, products, and services
had not changed significantly over the last year, James Smith determined that the current
EMS was suitable, adequate and effective for the corporation.

Figure 19-5. QSBI's documented management review.

20
Putting It All Together

The previous chapters have presented a perspective of the challenges that lie ahead for an organization that plans to implement an environmental management system (EMS) that conforms to ISO 14001. This chapter will provide some general thoughts about several current topics related to the implementation process. These include a discussion about preplanning and scheduling of the implementation process, a review of the potential for integration of the ISO 14001 system with ISO 9000, and some general advice about what to expect during the registration process. Finally, the authors will summarize the key lessons presented throughout the various chapters. As appropriate, our hypothetical company, Quality Seat Belts, Inc. (QSBI), will be used to illustrate key points.

Preplanning and Scheduling

Preplanning and scheduling of the implementation process are mandatory if an organization wants to successfully implement the elements of ISO 14001. Preplanning begins with a gap analysis of the organization's existing EMS with the elements of the standard. Based on the gap analysis, the organization's next step is to develop an implementation plan and schedule for closing the gaps. A gap analysis methodology used by QSBI is presented element by element in the appropriate chapters of this book. QSBI's implementation plan and 18-month schedule for implementing ISO 14001 and becoming registered to the standard is presented in Fig. 20-1.

Activity	Responsibility	F	M	A	M	J	J	A	S	O	N	D	J	F	M	A	M	J	J
Implementation Activities																			
Conduct gap analysis	Corporate	X																	
Develop implementation plan	Corporate		X																
Issue letter of support for the plan from president	Corporate		X																
Review and revise environmental policy	Corporate		X																
Communicate environmental policy	Corporate/ Plant sites		X	X															
Develop procedure for identifying aspects	Corporate			X															
Identify significant environmental aspects	Corporate/ Plant sites			X															
Develop procedure for legal/other requirements	Plant sites		X																
Set objectives and targets	Corporate/ Plant sites				X														
Develop environmental management program	Corporate/ Plant sites				X	X													
Develop structure and responsibility matrix	Corporate/ Plant sites					X													
Communicate roles and responsibilities	Corporate/ Plant sites					X	X												
Develop training and awareness procedure	Corporate					X													

Figure 20-1. QSBI's schedule for implementing the elements of ISO 14001.

Activity	Responsibility	F	M	A	M	J	J	A	S	O	N	D	J	F	M	A	M	J	J
Develop training and awareness presentations	Corporate/Plant sites						X	X											
Provide training to managers and employees	Corporate/Plant sites							X	X	X									
Update procedure for external communication	Corporate								X										
Develop procedures for internal communication	Corporate/Plant sites			X															
Formalize core elements of EMS	Corporate									X									
Develop document control procedure	Corporate				X														
Identify operations which warrant control	Corporate/Plant sites				X														
Develop/revise procedures for operational control	Corporate/Plant sites				X	X													
Develop procedure for operational control of contractors	Corporate/Plant sites				X						X								
Implement process for contractor awareness	Corporate/Plant sites					X	X												
Review/revise emergency response procedures	Plant sites				X														
Develop procedures for monitoring and measurment	Corporate/Plant sites				X														

Figure 20-1. (*Continued*) QSBI's schedule for implementing the elements of ISO 14001.

Activity	Responsibility	F	M	A	M	J	J	A	S	O	N	D	J	F	M	A	M	J	J
Implement procedures for monitoring and measurment	Plant sites					X	X	X											
Review/revise equipment calibration procedure	Plant sites					X													
Develop procedures for evaluation of compliance	Plant sites				X														
Develop procedure for handling nonconformances	Corporate					X													
Develop procedure for records retention	Corporate					X													
Develop EMS audit procedures and program	Corporate										X	X							
Perform EMS audits	Corporate												X	X	X				
Perform management review	Corporate/ Plant sites															X			
Begin continual improvement activities	Corporate/ Plant sites															X			
Registration Activities																			
Select registrar	Corporate												X						
Conduct preregistration meetings with registrar	Corporate/ Plant sites													X	X				

Figure 20-1. (*Continued*) QSBI's schedule for implementing the elements of ISO 14001.

Activity	Responsibility	F	M	A	M	J	J	A	S	O	N	D	J	F	M	A	M	J	J
Conduct initial registration audits at corporate and plant sites	Corporate/Plant sites															X	X		
Conduct final registration audits at corporate and plant sites	Corporate/Plant sites																X	X	
Close all nonconformances	Corporate/Plant sites																		X
QSBI registration certificates issued/plan celebration event	Registrar																		X

Figure 20-1. (*Continued*) QSBI's schedule for implementing the elements of ISO 14001.

The Potential for Integrating ISO 14001 with ISO 9000

Although this book has viewed ISO 14001 as a stand-alone system, there are many who advocate integrating ISO 14001 with ISO 9000. There are several elements that are core to each of these ISO standards. First, each is built around the "plan, do, check, and act" model. Second, each requires top-management commitment. After these elements, similarities begin to get somewhat "fuzzy," although proponents of joining the two systems think that they have much in common. An element-by-element comparison is presented in Table 20-1.

It should be noted that topic similarities should not be confused with similarities in substance. For instance, the policy statement in ISO 9000 is a statement of intent, whereas in ISO 14001, it is a statement of commitment. In terms of document control, ISO 9000 is very prescriptive, whereas ISO 14001 is purposefully not. Control of test equipment in ISO 9000 is, likewise, very detailed (if not onerous), whereas the equivalent of this requirement in ISO 14001—calibration of monitoring and measurement equipment—is very brief (one sentence) and very flexible. Finally, there is a fundamental difference between ISO 9000 and 14001 with respect to management review. With ISO 9000, the auditor makes the assessment if the quality system is effective during the audit process. With ISO 14001, only top management can make the assessment if the system is suitable, adequate, and effective, and audit results are only one element of this assessment. It is advisable for the organization to view its existing circumstances to ascertain advantages and disadvantages in integrating the two standards.

The Registration Process

After an organization has implemented ISO 14001, it may want to consider registration. If so, there are two registration methods allowed under the standard. The first is "self-declaration," in which the organization reviews its EMS in light of the requirements of ISO 14001 and deems that it is in conformance. The second, more stringent method, is through registration by an independent registrar accredited by a registration body. Certainly, for marketplace purposes, the latter means of accreditation is preferable.

Some general guidance about what might be reviewed during the registration audit—and which might not be intuitive—follows:

Table 20-1. ISO 14001 Requirements in Comparison with ISO 9000

ISO 14001 requirements	Equivalent ISO 9000 requirements	Comments about ISO 9000 requirements
General requirements	4.2.1	
Environmental policy	4.1.1	Quality policy
Planning		
Identification of environmental aspects	None	
Legal and other requirements	Addressed in ISO 9001, Section 4.4.4	
Objectives and targets	Addressed in ISO 9001, Section 4.1.1	
Environmental management program	None	
	4.2.3	Quality planning
Implementation and operation		
Structure and responsibility	4.1.2	Organization
Training, awareness, and competence	4.18	Training
Communication	None	
EMS documentation	4.2.1 (without 1st sentence)	General
Document control	4.5	Document and data control
Operational control	4.2.2, 4.3, 4.4, 4.6, 4.7, 4.9, 4.15, 4.19, 4.8	Control of various operations
Emergency preparedness and response	None	
Checking and corrective action		
Monitoring and measurement	4.10, 4.11, 4.12, 4.20	Inspection and testing
Nonconformance and corrective and preventive action	4.13, 4.14	Control of nonconforming product and corrective and preventive action
Records	4.16	Control of quality records
EMS audits	4.17	Internal quality audits
Management review	4.1.3	Management review

SOURCE: ISO 14001, Annex B (1996).

- *Environmental policy*—The organization should be able to demonstrate not only that it has a documented (and signed) environmental policy issued by top management but that its activities, products, and services support all the policy elements. Further, evidence that the policy has been communicated to relevant employees and contractors could be tested in the field through interviews.

- *Legal and other requirements*—Evidence that relevant employees have access to legal and other requirements may be requested. The organization should be able to demonstrate that voluntary requirements are adhered to as rigorously as legal requirements.

- *Environmental aspects*—The procedure for identifying environmental aspects could be reviewed to ensure that environmental aspects are identified first and that there is a process for determining which of these are significant. Evidence of how this process was deployed for each environmental aspect could be requested.

- *Objectives and targets/Environmental management program*—Evidence that significant environmental aspects were considered when setting objectives and targets could be requested. The means of achieving objectives and targets could be tested in the field.

- *Training, awareness, and competence*—Evidence that all relevant levels of employees—including top management—are aware of the EMS could be requested. That employees are aware of how their work activities could impact the environment and the objectives and targets (both favorably and negatively) would be tested through field interviews.

- *Communication*—Evidence that communication is multidirectional (i.e., mechanisms are provided at each relevant level for information sharing and feedback) may be requested. Evidence that procedures for handling external communications are being followed could be requested.

- *EMS documentation*—The organization should have clear links between the core elements and those that interact with these elements.

- *Document control*—The organization should carefully define which documents it wants to control, since these will likely be audited for proper authorization, dates, currency, and other requirements.

- *Operational control*—The organization should include maintenance in its operational control procedures. Evidence that relevant employees are trained in these procedures could be tested.

- *Emergency preparedness and response*—The organization should have a rationale for how and when to test its emergency procedures (or not to).

- *Monitoring and measurement*—The organization will need to determine what it considers its key characteristics—which should include at least the significant environmental aspects and possibly others. Evidence of these key characteristics and monitoring and measurement methods and data could be requested.

- *Nonconformance and corrective and preventive action*—Evidence that nonconformances are handled properly and that corrective and preventive actions are completed could be requested.

- *Records*—The organization should have these records identified and easily located, and they should be traceable to the activity, product, or service.

- *EMS audit*—In addition to EMS audit results, the rationale for scope and frequency of the audit could be requested.

- *Management review*—Evidence that the EMS was evaluated to ensure its suitability, adequacy, and effectiveness and that recommendations for continual improvement were made by top management could be requested.

Lessons Learned

Lessons learned from the discussions throughout this book are summarized in Fig. 20-2. In essence, ISO 14001 provides an organization with the tools it needs to systematize its EMS process and integrate it into the culture of the organization at all relevant functions and levels. The ISO 14001 framework provides the tools needed for committed organizations to improve their environmental management systems and, ultimately, environmental performance.

References

International Organization for Standardization, ISO 14001: Environmental Management Systems—Specification with Guidance for Use, Annex B, 1996.

Lessons Learned from Implementing ISO 14001

✓ Commitment to the EMS begins at the top.

✓ The EMS should be system dependent and not person dependent.

✓ The EMS is like a pyramid; the elements build upon one another.

✓ There needs to be a person who is responsible for the overall management of the EMS.

✓ Don't get caught in the "churn" when identifying significant environmental aspects.

✓ Use environmental processes and procedures that currently exist and align them with ISO 14001.

✓ The EMS does not have to be perfect to begin with; improvements can be made over time.

✓ Say what you do and do what you say, then have evidence to prove it.

✓ Communication and training are keys to success of the EMS.

✓ Every person in the organization can play a role in making the EMS stronger.

✓ Watch out for Section 4.4.6 (operational control); it is the ISO 14001 element most likely to cause nonconformance.

✓ An EMS audit is not the same as a compliance audit.

Figure 20-2. Lessons learned from implementing ISO 14001.

ISO 14001 Training Overheads

ISO 14001 Training Overheads

ISO 14001:
International Environmental
Management System Standard

Employee Training

What is ISO 14001?

- A voluntary international
  standard developed by the
  International Organization
  for Standardization

- It sets the requirements for the
  establishment of an environmental
  management system (EMS)

What are the Key Elements of ISO 14001?

- Environmental Policy
- Planning
- Implementation and Operation
- Checking and Corrective Action
- Management Review

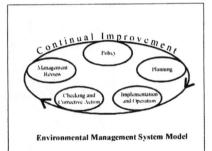

Environmental Management System Model

General Background

ISO 14001 Training Overheads

Key Benefits of ISO 14001

- Marketplace leverage
- Provides integrated approach
  to environmental management
- Is system dependent and not
  person dependent
- Demonstrates environmental
  leadership
- Promotes sound environmental
  management

Required Procedures

- Identification of significant
  environmental aspects
- Identification of legal and
  other requirements
- Internal and external communication
- Document control
- Operational control
- Emergency preparedness
  and response

Required Procedures, *Cont.*

- Monitoring and measurement of key
  characteristics
- Equipment calibration
- Evaluating of legal and regulatory
  compliance
- Maintaining records
- EMS audit

Keys to Success

- Management commitment
- Assigned responsibilities
- Implementation plan
- Education
- Appropriate procedures
  and documentation
- Audit readiness

General Background

ISO 14001 Training Overheads

Definition of Environmental Policy	Definition of Environmental Policy, *Cont.*
- Top management defines policy - Must be relevant to activities, products, and services - Must show commitment to continual improvement - Must show commitment to prevention of pollution	- Must show commitment to comply with legal and other requirements to which the organization subscribes - Must be communicated to employees - Must be available to the public
Example Environmental Policy Commitments	**Example Environmental Policy Commitments, *Cont.***
- Fulfill the responsibility of trustee of the environment for this and future generations - To the extent possible, apply practices and control technologies that minimize pollution	- Comply with current regulations worldwide - Strive for improvement of the environmental management system - Prevent pollution through reuse, recycling, and reduction

Environmental Policy

ISO 14001 Training Overheads

Identification of Environmental Aspects

- Significant environmental aspects of activities, products, and services must be defined
- A significant environmental aspect is one which has a significant environmental impact
- The environmental impact can be adverse or beneficial

Examples of Environmental Aspects

- Planned releases to air, water, and/or land
- Unplanned releases to air, water, and/or land
- Contamination of land
- Consumption of raw materials

Examples of Environmental Aspects, *Cont.*

- Consumption of natural resources
- Generation of waste
- Emission of heat
- Waste avoidance
- Creation of wildlife habitat

Significant Environmental Aspects

- These typically have the greatest environmental impact
- They reflect legal and/or regulatory requirements
- They are considered when setting objectives and targets

Planning

ISO 14001 Training Overheads

Identify Legal and Other Requirements
- Put in place procedures for identifying legal and other requirements to which the organization subscribes - Ensure relevant employees have access to these

Set Objectives and Targets
-These should be measurable if at all possible - These should support the environmental policy - These should consider the situation of the organization

Environmental Management Program
- This is a documented program - It gives the specifics of how the objectives and targets are to be achieved

Planning
- This section, along with the environmental policy, are the "heart" of the EMS - "Fail to plan and plan to fail"

Planning

ISO 14001 Training Overheads

Implementation and Operation

- Structure and responsibility
- Training, awareness, and competence
- Communication
- EMS documentation
- Document control
- Operational control
- Emergency preparedness and response

Structure and Responsibility

- Roles, responsibility, and authorities for establishing, implementing, and maintaining the EMS must be defined

- These roles, responsibility, and authorities must be communicated to relevant personnel

Training, Awareness, and Competence

- Employees must be aware of the EMS and its function
- Employees must understand how their work activities support the EMS and the environmental policy
- Employees must be competent in their job functions

Communication

- Information about the significant environmental aspects and the EMS must be communicated internally

- The organization must have a procedure for receiving, documenting, and responding to external communications

Implementation and Operation

ISO 14001 Training Overheads

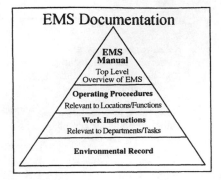

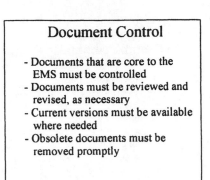

EMS Documentation

EMS
Manual
Top Level
Overview of EMS

Operating Proceedures
Relevant to Locations/Functions

Work Instructions
Relevant to Departments/Tasks

Environmental Record

Document Control

- Documents that are core to the
  EMS must be controlled
- Documents must be reviewed and
  revised, as necessary
- Current versions must be available
  where needed
- Obsolete documents must be
  removed promptly

Operational Control

- Includes control of operations
  related to significant environmental
  aspects and objectives and targets

- Includes control of on-site
  activities of contractors

- Procedures are required that specify
  operating criteria

**Emergency Preparedness
and Response**

- Procedures must be in place

- Procedures must be tested, as
  appropriate

Implementation and Operation

ISO 14001 Training Overheads

Monitoring and Measurement - Key characteristics must be monitored and measured to evaluate environmental performance - There must be a calibration schedule for monitoring and measuring equipment - Compliance with legal requirements must be evaluated	**Nonconformance and Corrective/Preventive Action** - Procedures must be in place to handle nonconformance to the EMS - Roles, responsibilities, and authorities must be assigned
Records - Environmental records must be easily located and retrievable - They must be legible, identifiable, and traceable to the activity, product or services - They must be protected against damage and loss	**EMS Audit** - An EMS audit is not a compliance audit - It should review the system's elements in light of planned arrangements, and should determine whether or not the EMS has been properly implemented and maintained

Checking and Corrective Action

ISO 14001 Training Overheads

Management Review

- Must be performed by top
  management

- EMS is reviewed to ensure its
  continuing suitability, adequacy,
  and effectiveness

Management Review

- Review process provides
  information about objectives
  and targets, monitoring and
  measurement results, EMS audit
  results, and other information

Management Review

- In light of the information presented,
  top management makes
  recommendations for changes to
  the EMS

- These recommended changes could
  be changes to policy, objectives and
  targets, or other elements of the EMS

Management Review

- Management's recommendations
  for changes should reflect the
  organization's commitment
  to continual improvement

Management Review

Environmental Management System Audit Methodology

EMS Audit Methodology

ISO 14001 Requirements	Audit Questions	Sources of Objective Evidence
4.1 General Requirements		
The organization must establish and maintain an environmental management system (EMS) that conforms to the requirements outlined in ISO 14001.	How has the EMS been established? How is it maintained?	Review elements of the EMS to ascertain if the EMS has been established and maintained in conformance with ISO 14001.

ISO 14001 Requirements	Audit Questions	Sources of Objective Evidence
4.2 Environmental Policy		
Top management must define the organization's environmental policy.	Does the organization have an environmental policy? Who issued the policy? How is top management defined?	Documented environmental policy. Evidence that it was issued by a top-management representative.
The policy must be appropriate to the nature, scale, and environmental impacts of the organization's activities, products, and services.	What are the activities, products, and services of the organization? How did the organization determine that the policy was appropriate to the nature, scale, and environmental impacts of its activities, products, and services?	Discussions with professionals about the organization's activities, products, and services. Tour of the facility to view activities, products, and services. Policy statements that address the environmental impacts and consequences of the organization's activities, products, and services.
The policy must include a commitment to continual improvement and prevention of pollution.	Does the policy include a commitment to continual improvement and prevention of pollution?	Statement in policy about commitment to continual improvement of the environmental management system and/or performance.
The policy must provide the framework for setting and reviewing environmental objectives and targets.	How does the policy provide the framework for setting and reviewing environmental objectives and targets?	Documented policy objectives plus additional documents such as environmental procedures, instructions, practices, guidance manuals, and/or other documents that build upon policy objectives.

ISO 14001 Requirements	Audit Questions	Sources of Objective Evidence
4.2 Environmental Policy (Cont.)		
The policy must be documented, implemented, and maintained and communicated to all employees.	Is the policy documented?	Documented policy.
	How is the policy implemented?	Evidence of implementation of each element of the policy.
	How is the policy communicated to all employees?	Interviews with employees to ascertain whether they know about the policy and how it relates to their jobs. Evidence that the policy has been communicated such as bulletin board campaigns, tent cards, and other information campaigns.
The policy must be available to the public.	How is the environmental policy made available to the public?	Evidence that the policy was made available to the public, including public announcements, publication in annual reports, posting of policy, new employee training, employee and neighborhood meetings, responses to public inquiries or requests, press clippings, distribution to contractors, or other evidence that this element has been met.

ISO 14001 Requirements	Audit Questions	Sources of Objective Evidence
4.3 Planning		
4.3.1 *Environmental aspects*		
The organization must establish and maintain a procedure to identify the environmental aspects of its activities, products, or services that it can control and over which it can be expected to have an influence. The organization must determine which aspects have or can have significant impacts on the environment. (These are termed significant environmental aspects.)	Has the organization established and maintained a procedure to identify the environmental aspects and impacts of its activities, products, and services?	Documented procedure(s). Interviews with persons responsible for developing, implementing, and maintaining the procedures. Evidence that the procedure(s) has been implemented and maintained such as a listing of significant environmental aspects.
The organization must ensure that the significant environmental aspects are considered in setting its environmental objectives.	Has the organization considered significant environmental aspects when setting its environmental objectives?	Evidence that significant environmental aspects are either part of the organization's environmental objectives or have been considered. Interviews with persons who set environmental objectives to verify that significant environmental aspects were considered. Meeting minutes of the process of setting environmental objectives.
The organization shall keep this information up to date.	Are the significant environmental aspects kept up to date?	Evidence that the identified list of aspects with significant environmental impact is up to date, such as last review or revision date on record. Interviews with personnel regarding the process of updating significant environmental aspects.

ISO 14001 Requirements	Audit Questions	Sources of Objective Evidence
4.3.2 Legal and other requirements		
The organization must have a procedure to identify and have access to legal and other requirements to which the organization subscribes which are applicable to the environmental aspects of its activities, products, or services.	Does the organization have a documented procedure to identify legal and other requirements to which the organization subscribes?	A documented procedure—which should include mention of international, national, regional, state or provincial, and local regulations; permit requirements; contracts or other documents that may have binding legal obligations. A list of voluntary requirements to which the organization subscribes such as national or state programs, industry codes of practice, and internal requirements. Interviews with environmental professionals about how legal requirements are accessed. Copies (hard copy or soft copy) of documents specifying legal and other requirements.
	How is access to legal requirements provided to those who need to know?	Procedures, work instructions, copies of the regulatory requirements, area postings of requirements. Interviews with employees who need access to legal requirements.

ISO 14001 Requirements	Audit Questions	Sources of Objective Evidence
4.3.3 Objectives and targets		
The organization must establish and maintain documented objectives and targets at relevant levels within the organization.	Are objectives and targets established and current? Are they set at relevant levels?	List of objectives and targets. Dates of review and/or update. Review of levels within the organization. Interviews with personnel at the various levels to ascertain if they are aware of the objectives and targets as they pertain to their jobs.
Several things must be considered when the organization sets its objectives and targets. These include considering legal and other requirements, significant environmental aspects, technological options, financial, operation and business requirements, and the views of interested parties.	Were the listed elements considered when the organization set its objectives and targets?	Meeting minutes of the process for setting objectives and targets. Interviews with personnel who set objectives and targets. Evidence of communication between relevant functions.

ISO 14001 Requirements	Audit Questions	Sources of Objective Evidence
4.3.4 Environmental management program(s)		
The organization must establish and maintain (a) program(s) for achieving its objectives and targets.	Does the organization have an environmental management program?	Documented program. Interview with personnel about the environmental management program.
The environmental management program must designate responsibility for achieving objectives and targets at each relevant function and level.	Are responsibilities designated at each relevant function and level of the organization for achieving objectives and targets?	Documented program. Organization charts. Job descriptions. Other documentation that shows responsibilities.
The environmental program must specify the means and time frame by which objectives and targets are to be achieved.	What are the means and time frames by which the objectives and targets are to be achieved?	Documented program. Program plans and schedules. Capital plans. Resource commitments. Other documentation of means and time frames.
If a project relates to new development and new or modified activities, products, or services, the program(s) must be amended where relevant to ensure that environmental management applies to such projects.	Is the environmental program amended when a project relates to new development and new or modified activities, products, or services, as necessary?	Date of documented program. Documentation pertaining to new projects. Interviews with project personnel.

ISO 14001 Requirements	Audit Questions	Sources of Objective Evidence
4.4 Implementation and Operation		
4.4.1 Structure and responsibility		
Roles, responsibility, and authorities must be defined, documented, and communicated.	Are roles, responsibility, and authorities defined and documented?	Defining of roles, responsibilities, and authorities through organization charts, charts detailing employees' responsibilities, job descriptions, and/or other documents.
	Are roles, responsibility, and authorities communicated appropriately?	Communication of roles through training classes, department meetings, management news bulletins, manager-employee meetings, other methods.
Management must provide resources essential to the implementation and control of the environmental management system. Resources include human resources and specialized skills, technology, and financial resources.	Has management provided appropriate: Human resources?	Headcount allocations for environmental management, including environmental professional staff, operations staff, and line staff.
	Financial resources?	Capital and expense plans for environmental management.
	Technology resources?	Plans for technology upgrades.

ISO 14001 Requirements	Audit Questions	Sources of Objective Evidence
4.4.1 Structure and responsibility (Cont.)		
The organization's top management must appoint (a) specific representative(s) who, irrespective of other responsibilities, shall have defined roles, responsibilities, and authority for: a) Ensuring that EMS requirements are established, implemented, and maintained in accordance to ISO 14001. b) Reporting on the performance of the environmental management system to top management for review and as a basis for improvement of the EMS.	Is there an EMS coordinator(s) or other representative(s) appointed by top management?	Letters of delegation and/or designation of an EMS coordinator(s) or other identified representative(s). Reports to management and others within the organization about EMS. Meeting minutes of management reviews.
4.4.2 Training, awareness, and competence		
The organization must identify training needs and require all personnel whose work may create a significant impact upon the environment to receive appropriate training.	Have training needs been identified?	Training plans or other documentation identifying who should receive training and what type. Interview with persons who specify training requirements.
	Have all employees whose work may create a significant impact upon the environment received appropriate training?	Training records to verify that training has been conducted, including classroom and on-the-job training.

ISO 14001 Requirements	Audit Questions	Sources of Objective Evidence
4.4.2 Training, awareness, and competence (Cont.)		
The organization must have procedures to make employees at each relevant function and level aware of the following: a) The importance of conformance with the environmental policy and procedures and with the requirements of the EMS. b) The significant environmental impacts, actual or potential, of their work activities and the environmental benefits of improved personal performance. c) Their roles and responsibilities in achieving conformance with the environmental policy and procedures and with the requirements of the EMS including emergency preparedness and response requirements. d) The potential consequences of departure from specified operating procedures.	Has the organization established procedures for employee awareness?	Training program to meet requirements of this element of the standard. Training rosters, meeting minutes, and other training records. Interviews with employees to assess awareness of EMS and operational activities in their job that could affect the environment.
Personnel performing the tasks which can cause significant environmental impacts must be competent on the basis of appropriate education, training, and/or experience.	Are personnel performing tasks which can cause significant environmental impacts competent on the basis of appropriate education, training, and/or experience?	Documents defining competency levels for employees performing tasks which can cause significant environmental impacts. Job descriptions which identify education levels and/or experience. Evidence that competency has been achieved through training records, job experience, or other means. Interviews with employees about adequacy of training. Interviews with management regarding employee competence.

ISO 14001 Requirements	Audit Questions	Sources of Objective Evidence
4.4.3 Communication		
The organization must have current procedures for: a) Internal communication between the various levels and functions of the organization. b) Receiving, documenting, and responding to relevant communication from external interested parties regarding its environmental aspects and EMS.	Are procedures established and maintained, as required?	Communications procedures that meet requirements of 4.4.3 a and b. Evidence of communications described in procedures for internal communications at each relevant function and level such as existence of distribution lists, hard copy and/or communication records, bulletin board announcements, and other communication records. Evidence of communications described in procedures for receiving, documenting, and responding to relevant communication from external interested parties including lists of calls, copies of letters, newspaper announcements, environmental reports, other external communications.
The organization must consider processes for external communication of its significant environmental aspects. The organization must record its decision on whether or not this external communication will take place and, if so, by what process.	Has the organization considered processes for external communication on its significant environmental aspects? Is the decision recorded?	Documented processes for communication with interested parties with respect to its significant environmental aspects. Interviews with external communication focal point(s) as to what processes were considered. Document of record about whether or not the organization will communicate its significant environmental aspects externally.

ISO 14001 Requirements	Audit Questions	Sources of Objective Evidence
4.4.4 Environmental management system documentation		
The organization must describe the core elements of its EMS and their interaction and provide direction to related documentation. EMS documentation can be in paper or electronic form.	Has the organization established information to describe the core elements of the EMS and to provide direction to related documentation? Is the information maintained?	Documented EMS, including the core elements and their interaction. Dates of EMS updates/reviews. List or index of related documentation. Interviews with environmental professionals about what makes up the EMS and related documentation.
4.4.5 Document control		
The organization must establish and maintain procedures for controlling all documents required by ISO to ensure that: a) They can be located. b) They are periodically reviewed, revised as necessary, and approved for adequacy by authorized personnel. c) The current versions of relevant documents are available at all locations where operations essential to the effective functioning of the system are performed. d) Obsolete documents are promptly removed from all points of issue and points of use or otherwise assured against unintended use. e) Any obsolete documents retained for legal and/or knowledge preservation purposes are suitably identified.	Are procedures available to ensure appropriate document control?	Document control procedures. Records of document review and revision. Evidence that procedures are met including review of point-of-use areas for current documents.

ISO 14001 Requirements	Audit Questions	Sources of Objective Evidence
4.4.5 Document control (Cont.)		
Documentation shall be legible, dated (with dates of revision), and readily identifiable, maintained in an orderly manner, and retained for a specified period.	Is documentation legible, dated, and readily identifiable, maintained in an orderly manner, and retained for a specified period?	Evidence through document sampling.
Procedures and responsibilities shall be established and maintained concerning the creation and modification of various types of documents.	Are procedures and responsibilities established and maintained concerning the creation and modification of various types of documents?	Procedures and documented responsibilities. Interviews with persons responsible for maintaining document control. List of controlled documents.
4.4.6 Operational control		
The organization must identify those operations and activities that are associated with the identified significant environmental aspects.	Has the organization identified those operations and activities that are associated with the identified significant environmental aspects?	Documented operations and activities that are associated with significant environmental aspects. Interviews with personnel associated with these operations and activities.

ISO 14001 Requirements	Audit Questions	Sources of Objective Evidence
4.4.6 Operational control (Cont.)		
With respect to these activities, the organization must: a) Establish and maintain documented procedures to cover situations where their absence could lead to deviations from the policy and the objectives and targets. b) Stipulate operating criteria in the procedures. c) Establish and maintain procedures related to the identifiable significant environmental aspects of goods and services used by the organization. The organization must communicate relevant procedures and requirements to suppliers and contractors.	Are procedures documented and maintained for relevant activities, including maintenance, to cover situations where their absence could lead to deviations from the policy and objectives and targets? Are operating criteria stipulated in the procedures? Are procedures established and maintained related to the identifiable significant environmental aspects of goods and services used by the organization? Are relevant procedures and requirements communicated to suppliers and contractors?	Procedures; implementation of procedures; date of revision/review of the procedures. Interviews with personnel who carry out the procedures. Procedures, including operating criteria. Interview with personnel who carry out the procedures. Procedures related to the identifiable significant environmental aspects of goods and services used by the organization; date of revision/review of these procedures. Evidence of communication of procedures and requirements to suppliers and contractors, including purchasing contracts, contractor meetings, contractor guidance brochures, other.

ISO 14001 Requirements	Audit Questions	Sources of Objective Evidence
4.4.7 Emergency preparedness and response		
The organization must establish and maintain procedures to identify potential for and respond to accidents and emergency situations, and for preventing and mitigating the environmental impacts that may be associated with them.	Has the organization established and maintained procedures to identify potential for and response to accidents and emergency situations and for preventing and mitigating the environmental impacts that may be associated with them?	Procedures for identifying potential for accidents and date of revision/review. Process hazard assessments and other risk assessments. Procedures for response to accidents and emergency situations and for preventing and mitigating the environmental impacts that may be associated with them and dates of revision/review.
The organization must review and revise, where necessary, its emergency preparedness and response procedures and, in particular, after the occurrence of accidents or emergency situations.	Has the organization reviewed and revised, where necessary, its emergency preparedness and response procedures after the occurrence of accidents or emergency situations?	Records of actions taken in response to and following emergency or accident situations. Interviews with personnel who participated with review of procedures or emergency response activities.
The organization shall periodically test such procedures where practicable.	Has the organization periodically tested emergency preparedness and response procedures?	Documentation of test drills, table-top scenarios, and other methods of testing emergency preparedness and response. Documentation of changes in procedures as a result of such tests.

ISO 14001 Requirements	Audit Questions	Sources of Objective Evidence
4.5 Checking and Corrective Action		
4.5.1 Monitoring and measurement		
The organization must have documented procedures to monitor and measure on a regular basis the key characteristics of its operations and activities that can have a significant impact on the environment. Information must be recorded to track performance, relevant operational controls, and conformance with the organization's objectives and targets.	Has the organization identified key characteristics? Has the organization established and maintained procedures to monitor and measure on a regular basis the key characteristics of its operations and activities? Does the procedure include the recording of information to track performance, relevant operational controls, and conformance with the organization's objectives and targets?	Documented procedure and dates of revisions/review. Evidence of implementation of procedure including monitoring and measurement data. Interviews with person who collects the monitoring and measurement data.
Monitoring equipment must be calibrated and maintained and records of this process must be retained, per the procedures.	Are there procedures defining a calibration schedule for monitoring equipment? Is monitoring equipment calibrated and maintained and are records of this process retained according to the organization's procedures?	Procedures and dates of review/revision. List of equipment that is controlled under the procedure. Calibration schedules and records for this equipment. Interviews with persons who perform the calibration.
The organization must have a documented procedure for periodically evaluating compliance with relevant environmental legislation and regulations.	Has the organization established and maintained a documented procedure for periodically evaluating compliance with relevant environmental legislation and regulations?	Documented procedure and date of revision/review. Evidence of compliance self-assessments or field checks, such as audit records and reports. Schedule for performing such evaluations. Interviews with personnel who conduct the compliance assessments.

ISO 14001 Requirements	Audit Questions	Sources of Objective Evidence
4.5.2 Nonconformance and corrective and preventive action		
The organization must have procedures for defining responsibility and authority for handling and investigating nonconformance, taking action to mitigate any impacts caused, and initiating and completing corrective and preventive action.	Has the organization established and maintained procedures for defining responsibility and authority for handling and investigating nonconformance, taking action to mitigate any impacts caused, and initiating and completing corrective and preventive action?	Procedures and date of revision/review. Documented list or other evidence of nonconformances to the EMS and/or environmental performance, corrective actions, and preventive actions. Tracking documents that provide evidence that the corrective and preventive actions were implemented. Records of field checks of corrective actions.
Any corrective or preventive action taken to eliminate the causes of actual or potential nonconformance must be appropriate to the magnitude of problems and commensurate with the environmental impact encountered.	Are any corrective or preventive actions taken to eliminate the causes of actual or potential nonconformance appropriate to the magnitude of problems and commensurate with the environmental impacts encountered?	Documentation of corrective or preventive actions. Evidence that the corrective or preventive actions have been evaluated as adequate and appropriate for the magnitude of the problem(s) such as meeting minutes. Interviews with persons responsible for developing and/or implementing corrective actions.
The organization must implement and record any changes in documented procedures resulting from corrective and preventive action.	Has the organization implemented and recorded any changes in documented procedures resulting from corrective and preventive action?	Procedures associated with this element and date of review/revision. Interviews with persons performing work using the new procedures.

ISO 14001 Requirements	Audit Questions	Sources of Objective Evidence
4.5.3 Records		
The organization must have procedures for the identification, maintenance, and disposition of environmental records. These records must include, at a minimum, training records and the results of audits and reviews.	Has the organization established and maintained procedures for the identification, maintenance, and disposition of environmental records?	Procedures and date of revision/review. List of environmental records and associated disposition.
	Do the records include training records and results of audits and reviews?	Training records, audit records, management review records, and other records identified by the procedure. Interviews with person(s) responsible for maintaining the records.
Environmental records must be legible, identifiable, and traceable to the activity, product, or service involved. Environmental records must be stored and maintained in such a way that they are readily retrievable and protected against damage, deterioration, or loss. Their retention times must be established and recorded.	Are the organization's environmental records legible, identifiable, and traceable to the activity, product, or service involved?	Environmental records as they related to the activity, product, or service involved.
	Are the organization's environmental records stored and maintained in such a way that they are readily retrievable and protected against damage, deterioration, or loss?	Procedure (if available) for storage and maintenance of records. Evidence of implementation of procedure or process that allows for retrievability, protection against damage, deterioration, and loss through sampling of records. Physical condition of records.
	Are there established and recorded retention times for these records?	List of record retention times. Sampling of records to assess retention times.
Records must be maintained to the requirements of this section.	Are records maintained to the requirements of this section?	Evidence that records are maintained appropriately through sampling of records.

ISO 14001 Requirements	Audit Questions	Sources of Objective Evidence
4.5.4 Environmental management system audit		

The organization must establish and maintain (a) program(s) and procedures for periodic EMS audits to be carried out, in order to:
 	
a) Determine whether or not the environmental management system
 	
   1) Conforms to planned arrangement for environment including the requirements of this standard.
 	
   2) Has been properly implemented.
 	
b) Provide information on the results of audits to management.
	Has the organization established and maintained a program and procedures for periodic environmental management system audits?
	
	Does the program determine whether or not the environmental management system conforms to planned arrangements for environmental management including the requirements of this standard and whether or not it has been properly implemented?
	
	Does the program and procedures provide information on the results of audits to management?
		Program and procedures and date of revision/review. Audit checklist. Audit schedules. Evidence of information from audit that is to be reviewed with management such as meeting minutes or presentation materials.

The audit program, including any schedule, shall be based on the environmental importance of the activity concerned and the results of previous audits.	Is the audit program, including any schedule, based on the environmental importance of the activity concerned and the results of previous audits?	Audit schedule. Audit results. Interviews with personnel responsible for audit schedule. Revisions to schedule based on findings, as appropriate.
In order to be comprehensive, the audit procedures must cover the audit scope, frequency, and methodologies, as well as the responsibilities and requirements for conducting audits and reporting results.	Do the audit procedures cover the audit scope, frequency, and methodologies, as well as the responsibilities and requirements for conducting audits and reporting results?	Procedures and date of revision/review. Audit schedules. Audit checklist or other methodologies. Audit responsibilities including conducting audits and reporting results. Interviews with persons who conduct the EMS audits.

ISO 14001 Requirements	Audit Questions	Sources of Objective Evidence
4.6 Management Review		
The organization's top management must, at interval it determines, review the EMS, to ensure its continuing suitability, adequacy, and effectiveness.	Does the organization's top management, at interval it determines, review the EMS, to ensure its continuing suitability, adequacy, and effectiveness?	Top management EMS review schedule. Evidence that top management has reviewed the EMS per schedule to ensure continuing suitability, adequacy, and effectiveness.
The management review process must ensure that the necessary information is collected to allow management to carry out the evaluation of the EMS.	Does the management review process ensure that the necessary information is collected to allow management to carry out this evaluation?	Management review process. Evidence of information that management is shown during reviews. Interviews with person(s) who present to top management.
This review must be documented.	Is the review documented?	Documentation of management reviews.
The management review must address the possible need for changes to policy, objectives, and other elements of the environmental management system, in light of environmental management system audit results, changing circumstances, and the commitment to continual improvement.	Does the management review address the possible need for changes to policy, objectives, and other elements of the environmental management system, in light of environmental management system audit results, changing circumstances, and the commitment to continual improvement?	Documentation of changes to EMS, as appropriate. Documentation of changing circumstances. Evidence of continual improvement of the EMS such as improvement in environmental performance, improvements to objectives and targets, and other improvements.

Index

About the Authors

GAYLE WOODSIDE is program manager on IBM's Corporate Environmental Affairs staff, working on ISO 14001 worldwide implementation. She is also the author of McGraw-Hill's *ISO 14000 Guide: The New International Environmental Management Standards* and the forthcoming *EHS Professional's Portable Handbook.*

PATRICK AURRICHIO works on IBM's Corporate Environmental Affairs staff, where he is responsible for ISO 14001 worldwide implementation, monitoring and measurement, and EMS training. He also provides technical support for property transactions, environmental assessments, and EMS auditing.

JEANNE YTURRI is vice president of Zephyr Environmental Corporation, an engineering consulting firm in Austin, Texas. She provides a variety of technical consulting services to clients, including environmental, safety and health program compliance audits, and ISO 14001 implementation assistance.